Protecting Group Chemistry

Jeremy Robertson

Lecturer in Organic Chemistry and Fellow of Brasenose College, Oxford

Series sponsor: AstraZeneca

AstraZeneca is one of the world's leading pharmaceutical companies with a strong research base. Its skill and innovative ideas in organic chemistry and bioscience create products designed to fight disease in seven key therapeutic areas: cancer, cardiovascular, central nervous system, gastrointestinal, infection, pain control, and respiratory.

AstraZeneca was formed through the merger of Astra AB of Sweden and Zeneca Group PLC of the UK. The company is headquartered in the UK with over 50,000 employees worldwide. R&D centres of excellence are in Sweden, the UK, and USA with R&D headquarters in Södertälje, Sweden.

AstraZeneca is committed to the support of education in chemistry and chemical engineering.

OXFORD
UNIVERSITY PRESS

OXFORD

UNIVERSITY PRESS

Great Clarendon Street, Oxford OX2 6DP
Oxford University Press is a department of the University of Oxford.
It furthers the University's objective of excellence in research, scholarship,
and education by publishing worldwide in

Oxford New York

Athens Auckland Bangkok Bogotá Buenos Aires Calcutta
Cape Town Chennai Dar es Salaam Delhi Florence Hong Kong Istanbul
Karachi Kuala Lumpur Madrid Melbourne Mexico City Mumbai
Nairobi Paris São Paolo Singapore Taipei Tokyo Toronto Warsaw

with associated companies in Berlin Ibadan

Oxford is a registered trade mark of Oxford University Press
in the UK and in certain other countries

Published in the United States
by Oxford University Press Inc., New York

A catalogue record for this book is available from the British Library

Library of Congress Cataloging in Publication Data
(Data applied for)

ISBN 0 19 850275 3

Typeset by the author
Printed in Great Britain
on acid-free paper by Bath Press Ltd, Bath, Avon

Series Editor's Foreword

Protecting groups are not only one of the most useful features of the synthetic armoury but they also provide an excellent forum for the teaching and understanding of basic mechanistic chemistry. Only by understanding the fundamental chemistry associated with the functional groups of each protecting group can successful orthogonal protecting group strategies be elucidated. The whole area of protecting group chemistry therefore provides an ideal fertile teaching area for basic organic chemistry.

Oxford Chemistry Primers have been designed to provide concise introductions relevant to all students of chemistry and contain only the essential material that would normally be covered in an 8–10 lecture course. This present Primer by Jeremy Robertson presents the concepts, mechanisms and applications of protecting groups in a unique, very logical and student friendly fashion. This Primer will be of interest to apprentice and master chemist alike.

Professor Stephen G. Davies
The Dyson Perrins Laboratory,
University of Oxford

Preface

Protecting groups are used to block the reactivity of specific functional groups within a molecule so that chemical transformations may be carried out elsewhere without ambiguity. For students approaching this subject for the first time there seems to be a vast array of seemingly unrelated protecting groups that are used in every conceivable combination for no apparent reason; it has been my aim in writing this book to try to convey the sense that, in fact, the vast majority of protecting groups may be divided into a mere handful of different categories and that members of each category can be removed under broadly similar experimental conditions.

The suitability of a particular protecting group in a given situation is dictated by the level of protection imparted and the conditions required to remove it when no longer required; this latter feature is, in turn, a function of the chemical reactivity of the components of the protecting group and the reaction pathways available to them. Therefore the coverage is organised not, as is more usual, around the functional groups themselves, but around the conditions that cause protecting group cleavage and the mechanistic principles that underlie this reactivity.

I would like to thank Caroline for her unending patience and support during the preparation of this book, Steve Davies for all the helpful comments on draft versions of the text, and the members of my research group who have had to endure my 'Primer' preoccupation in recent months.

Oxford   J. R.
May 2000

Contents

1   Introduction   1

2   Acid-labile protecting groups   18

3   Nucleophile/base-labile protecting groups   46

4   Silyl protecting groups   74

5   Redox deprotection   83

Index   95

1 Introduction

1.1 Introduction

The pharmaceutical and agrochemical industries undertake multistep syntheses of complex organic molecules that possess desirable biological properties; studies of biomolecules—proteins, oligosaccharides, and nucleic acids—and the rapidly evolving areas of supramolecular chemistry and molecular electronics also rely on sophisticated synthesis. As structural complexity increases, so does the need to control the diverse reactivity exhibited by a wide variety of functional groups within a molecule. This book will survey a central aspect of the strategy of molecular synthesis, i.e. methods that aid the selective modification of molecules at specific sites by masking or protecting any other parts of the molecule that could potentially interfere.

This sub-discipline of chemical synthesis—protecting group chemistry—has been the subject of a number of excellent books and reviews that have been written largely with the practising research chemist in mind; these usually emphasise available laboratory methods rather than the underlying chemistry that guides the choice of protecting group. The approach adopted here places protecting groups within a mechanistic framework in order to convey an appreciation of how a protecting group can be selected from the multitude available.

1.2 Strategy

In general, a protecting group (PG) is introduced onto a functional group (FG) to block its reactivity under the experimental conditions needed to make modifications elsewhere. The conditions under which a particular PG is either removed or left unaffected must be known; therefore, this book is organised around the mechanistic principles that influence PG reactivity.

For example, consider the selective oxidation of just the 3-hydroxyl group (starred in the scheme below) in glucose. It is unlikely that we could find a reagent to achieve this, so the other four hydroxyl groups must be prevented from reacting. In this specific case we can take advantage of the fact that glucose reacts with acetone in the presence of an acid catalyst to form a single compound in which the 3-hydroxyl group is unaffected. Oxidation with PDC is then unambiguous and subsequent acidic hydrolysis reveals the four hydroxyl groups. Overall, acetone protects four of the hydroxyl groups in glucose by forming two cyclic acetal derivatives (acetonides, Chapter 2).

This use of two acetonide PGs allows the ketone to be prepared in good yield but a negative aspect of PG chemistry is immediately apparent; two extra synthetic steps have to be added to the sequence, one to introduce the acetonide PGs and one to remove them after oxidation. This is somewhat

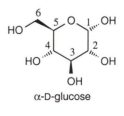

α-D-glucose

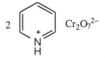

pyridinium dichromate (PDC)

inefficient but, until organic synthesis reaches the level where there is a reagent available for every eventuality, we are stuck with the need to use PGs and must concentrate on maximising their efficiency.

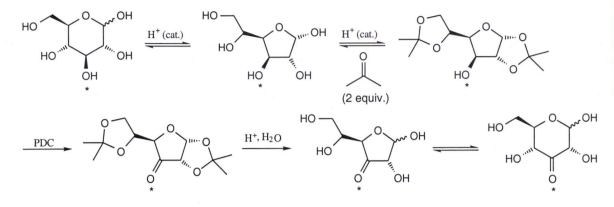

*Non-stereoselective introduction of a new stereogenic centre leads to the production of a mixture of diastereomers which complicates both purification and spectroscopic characterisation.

Given that PGs can be unavoidable, we have to place fairly stringent limitations on their properties if they are to be generally useful. Ideally, a PG should be introduced in quantitative yield at the desired site under conditions that cause no alteration of the rest of the molecule, preferably without introducing new stereogenic centres.* The PG must remain intact during all the subsequent reactions and must neither induce side-reactions nor affect the outcome of the chemistry in an unpredictable manner. Ultimately, when no longer required, the PG must be removable under specific conditions that cause no disruption of the rest of the molecule. Accommodating all these requirements (and others: cost, toxicity, etc.) is a tall order and very few PGs have proven to be generally applicable to a wide variety of situations; this is one reason why there are so many!

In the middle of a synthesis it is usually desirable to be able to remove specific PGs without cleaving others but at the end of a synthesis it is more efficient to remove all of the PGs in a single step. This leads to two generalisations that guide PG strategy: (1) Reaction conditions should be available that remove just one type of PG leaving all the other types intact and groups of these other types should, when necessary, be removable without affecting the first type. Groups of PGs that can be deprotected under these 'mutually exclusive' conditions are often referred to as *orthogonal sets*. For example, acetate esters (-OAc) are cleaved reliably with NH_3 but generally are unaffected by catalytic hydrogenolysis, whereas benzyl ethers (-OBn) are essentially inert to amine nucleophiles and catalytic hydrogenolysis is the method of choice for their removal.

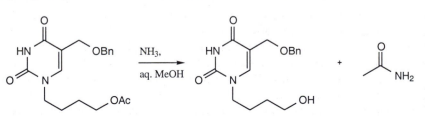

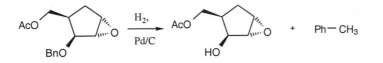

(2) Removing all the PGs at the end of the synthesis in a single step is incompatible with the presence of orthogonal sets of PGs; thus, within each set, there should be a degree of fine-tuning to allow selectivity during the synthesis ('graduation of lability'). Evans' synthesis of cytovaricin, an antibiotic, made heavy use of silyl PGs of varying stability to allow differentiation during the synthesis. However, in the final step, HF was added as it is sufficiently powerful to remove all seven silicon atoms in four types of silyl PGs to expose the eight hydroxyl groups in the natural product.

$t\text{-BuMe}_2\text{Si-}$	*tert*-butyldimethylsilyl (TBS)
$\text{Et}_3\text{Si-}$	triethylsilyl (TES)
$i\text{-PrEt}_2\text{Si-}$	diethylisopropylsilyl (DEIPS)

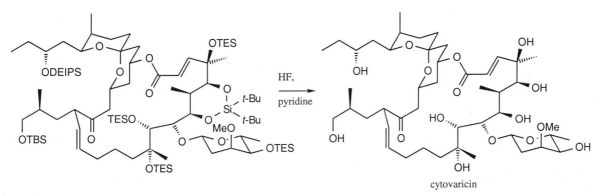

cytovaricin

1.3 Functional group reactivity

Broad strategic issues aside, the choice of PG is heavily influenced by the type of reactivity that must be blocked which requires a knowledge of the characteristic reactivity of each FG. Although this is discussed in detail in OCP #35 (see Further reading list on page 17), the key FG characteristics are summarised here in the context of their need for protection.

Alcohols (R–OH)

The lone electron pair(s) borne by heteroatoms—O, N, and S—in their lower oxidation states impart nucleophilicity (in general: O < N < S). Alcohols therefore react readily with oxidising agents (→ carbonyl compounds) and with other electrophiles (alkyl halides → ethers, acyl halides → esters); if electrophilic reagents are required and reactions at selected hydroxyl sites are to be avoided, this nucleophilic behaviour must be deactivated. Sterically encumbering the hydroxyl oxygen with a bulky PG can prevent it competing effectively for electrophiles. The triphenylmethyl (trityl, Tr) ether PG is an example; rotation about the O–CPh$_3$ bond is rapid and the phenyl rings sweep out a large volume around the oxygen atom to prevent the approach of anything but the smallest electrophiles (e.g. H$^+$, page 22).

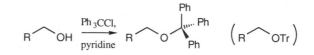

The inherent nucleophilicity of an oxygen atom may be reduced by delocalising the lone electron pairs by conjugation with, for example, a carbonyl π-system. For example, pivaloate (Pv) esters can be introduced selectively onto 1°-hydroxyl groups and have the added advantage that the carbonyl carbon is sterically hindered (by the *tert*-butyl group) towards attack by nucleophilic reagents.

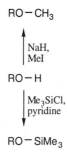

Delocalisation reduces nucleophilicity.

The second alkyl oxygen lone pair (not shown) resides in an sp^2 orbital orthogonal to the π-system.

RO–H   pK_a ≈ 15–17
(HO–H   pK_a = 15.74)

RO$-$CH$_3$

↑ NaH, MeI

RO$-$H

↓ Me$_3$SiCl, pyridine

RO$-$SiMe$_3$

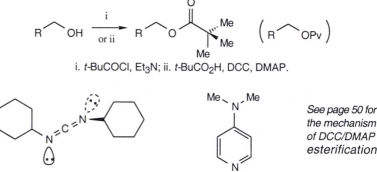

i. *t*-BuCOCl, Et$_3$N; ii. *t*-BuCO$_2$H, DCC, DMAP.

dicyclohexylcarbodiimide
(DCC)

4-(dimethylamino)pyridine
(DMAP)

See page 50 for the mechanism of DCC/DMAP esterification.

Hydroxyl groups possess a readily removed proton which interferes with the action of strongly basic reagents including Grignards (R–MgX) and other organometallics that are destroyed by protonation.

R–MgX + R′–OH → R–H + R′–OMgX

Replacement of the hydroxyl proton with a relatively inert stand-in avoids this problem. For example, alkyl ethers are easy to prepare (Williamson ether synthesis and modifications), they survive a wide variety of reaction conditions, but can be difficult to remove when no longer required. Modified alkyl ethers, that incorporate additional functionality, may be removed under mild conditions; examples of these PGs will appear throughout the course of the book. Silyl ethers are easily introduced with silyl chlorides or triflates (trifluoromethanesulfonates) and offer a number of advantages over alkyl ethers: (1) they reduce the effective nucleophilicity of the oxygen atom sterically and electronically, (2) the silyl alkyl substituents can be chosen to provide the desired level of protection, and (3) they may be readily removed under highly selective conditions.

Amines (R–NH₂, R–NHR′, R–NR′R″)

Proton removal from 1°- and 2°-amines by basic reagents is not as great a problem as it is with alcohols because the nitrogen atom has a reduced ability to bear a full negative charge (electronegativity: page 46). Even so, many organometallic reagents *are* sufficiently basic to fully deprotonate amines* and protection is required in these instances. More importantly, this weaker capability to support negative charge is manifested in a much higher nucleophilicity, e.g.

$R_2N–H \quad pK_a \approx 35$
$(H_2N–H \quad pK_a = 38)$

$i\text{-}Pr_2NH + BuLi \rightarrow i\text{-}Pr_2NLi$
*Preparation of lithium diisopropylamide (LDA).

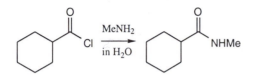

Water is present in large excess as the solvent but it cannot compete for the acyl chloride with the amine (which is much more nucleophilic).

Because amines compete *so* effectively for electrophiles it can be very difficult to handle molecules that bear free amines—both from the point of view of reactivity and ease of purification—therefore a good plan is to keep the amine in a protected form for as long as possible, only deprotecting it at the very end of the route. As with alcohols, the nucleophilicity of amines can be reduced by steric encumbrance but PGs that involve the lone pair of electrons in another bond are much more reliable. Quaternisation with a reactive alkyl halide (MeI, BnBr) or protonation to produce an ammonium salt completely remove the nucleophilicity of the amine; this finds use in protecting 3°-amines during oxidation.

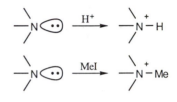

Alternatively, the nitrogen lone pair can be brought into conjugation with a π-system in analogy to the use of ester PGs for alcohols; many amides (e.g. formamides, acetamides, trifluoroacetamides) and carbamates have been devised that differ in their tolerance of nucleophiles and their deprotection requirements.

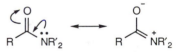

Carbamates (urethanes, R₂NCO₂R′) may be prepared easily by treating the amine with an alkyl chloroformate under Schotten–Baumann conditions (aq. NaOH) or with the appropriate anhydride and equivalent reagents. These derivatives impart excellent deactivation of the nitrogen's nucleophilicity and the carbonyl group shows good stability towards nucleophiles. Their main advantage is that the *O*-alkyl group can be chosen to allow mild and specific deprotection by methods that do not require nucleophilic addition to the carbonyl group.

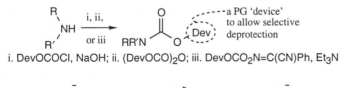

i. DevOCOCl, NaOH; ii. (DevOCO)₂O; iii. DevOCO₂N=C(CN)Ph, Et₃N

The carbamate C=O group is stabilised by mesomeric electron donation by both the N and O atoms and shows very low reactivity towards nucleophilic attack. However, the two heteroatoms 'compete' for the carbonyl so they are individually *less* involved in conjugation than in amides and esters respectively (a carbamate N is more basic than an amide N).

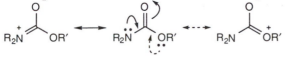

Thiols (R–SH)

The chemistry of sulfur compounds is discussed in detail in OCP #33 but in the context of PG chemistry the most important feature is that the valence electrons centred on sulfur are readily polarisable with the result that thiols and sulfides are highly nucleophilic and easily oxidised. Therefore, PG issues of sulfur-containing molecules parallel those of amines and many of the PGs are analogous acyl derivatives (thioesters, thiocarbonates, thiocarbamates) that will not be discussed separately.

In contrast to the N–H bond in amines, the S–H bond is weak and alkyl thiols are markedly acidic (conjugating S-substituents raise this acidity). Accordingly, thiol protection is required to prevent deprotonation by basic reagents and, for this purpose, PGs based on those used in alcohol protection are applied.

However, a characterising feature of thiol chemistry is ready oxidative dimerisation to give disulfides. Fortunately, this process is reversible and free thiols may be regenerated with a variety of reducing agents. In cases where the weak S–S bond is tolerated, thiol protection by disulfide formation may be the most convenient method.

RS–H $pK_a \approx 10$–11
PhS–H $pK_a = 6.5$

RS–H ≈ 365 kJ mol^{-1}
cf. RO–H ≈ 435 kJ mol^{-1}
RS–SR ≈ 290 kJ mol^{-1}
PhS–SPh ≈ 105 kJ mol^{-1}

Bond strengths in sulfur compounds.

Symmetrical disulfides

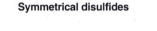

i. O$_2$ or H$_2$O$_2$ or I$_2$, etc.;
ii. NaBH$_4$ or Na or R'SH, etc.

Unsymmetrical disulfides

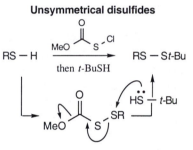

Carbonyl groups (R$_2$C=O, RCHO)

The ability of the carbonyl group to act both as an electrophile and as a nucleophile makes it of central importance in the formation of carbon–carbon bonds. The aldol reaction, effected with either basic or acidic catalysis, illustrates both aspects of carbonyl reactivity.

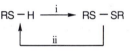

electrophilic

$pK_a \approx 20$

Carbonyl chemistry is characterised by nucleophilic attack at the C=O carbon and by enol(ate) formation followed by α-functionalisation.

Base catalysed

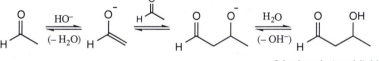

3-hydroxybutanal ('aldol')

Acid catalysed

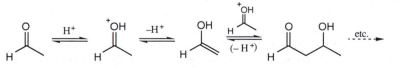

Base-catalysed enolate formation [or acid-catalysed enol formation] transforms the inherently electrophilic carbonyl group into a nucleophilic enolate anion [or enol] which is manifested in its addition to free aldehyde [or protonated aldehyde] to form the β-hydroxy aldehyde ('aldol') product. Because the aldol reaction proceeds readily in the presence of catalytic quantities of acid *or* base, its occurrence as a competing side-reaction is difficult to avoid and carbonyl protection is a central requirement for successful organic synthesis.

Fortunately, there are plenty of PGs that remove both aspects of carbonyl reactivity simultaneously. Most are based on alcohol adducts (acetals) formed under dehydrating conditions in the presence of protic or Lewis acids. The generic reaction, to form 1,1-diethoxyethane ('acetal'), is shown:

The term 'ketal' to distinguish ketone acetals from aldehyde acetals has been discarded by IUPAC.

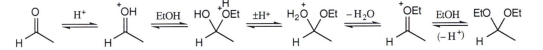

The acetal carbon (to which the alkoxy groups are attached) is less electron deficient, more sterically crowded, and no longer bears a polarised and relatively easily broken π-bond; it is therefore not susceptible to direct attack by nucleophiles. The lack of a potentially conjugating π-system reduces the acidity of the α-protons so effectively that only the very strongest bases will remove them. Thus acetals effectively shield the carbonyl group from both nucleophilic attack and base-induced enolate formation.

Acetal formation is, however, a reversible process with the equilibrium (usually) lying in favour of the carbonyl compound;* therefore, these PGs cannot be used reliably when acidic reagents are employed, especially if water is also present.

*Usually, the liberated water must be removed in order to drive the equilibrium to completion.

Carboxyl groups (R–CO$_2$H)

Proton removal from carboxylic acids presents a threat to potentially valuable organometallic reagents, resulting in an alkane and a metal carboxylate.

RCO$_2$–H   pK_a ≈ 3–5

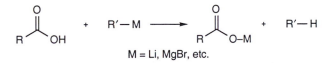

M = Li, MgBr, etc.

If the so-formed carboxylate leads to no undesirable secondary reactions and the organometallic reagent is readily available, then use of an excess of reagent can bypass this problem. Usually, neither of these requirements apply and suitable carboxyl PGs are needed. Furthermore, protection against enolisation or attack by nucleophiles may be necessary; few PGs meet all of these requirements. Ester and amide derivatives remove the problem of the carboxyl proton, and the latter provide good protection against many nucleophiles, but neither prevent enolisation by strong bases. A work-around is built on the reliable oxidation of phenyl groups by RuO$_4$ (prepared *in situ*) to generate carboxylic acids in the presence of diverse functionality.

This is not, strictly speaking, FG protection but FG interconversion of a type that is more within the scope of OCP #35.

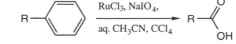

However, orthoesters provide complete protection of the carboxyl group. They can be prepared directly from the acid (although this is usually low yielding)

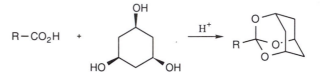

or via an oxetanylmethyl ester in a two-step procedure:

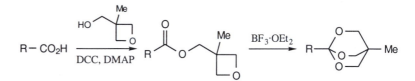

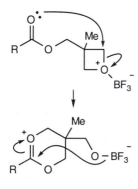

Carboxyl protection is required during the condensation of two amino acids with the aid of a coupling agent to form a new peptide bond (OCP #7). To prevent the formation of self-condensation products, one amino acid requires a free carboxyl group and a protected amino group, the other requires a protected carboxyl group and a free amino group.

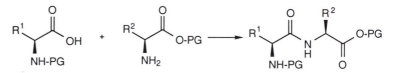

The carboxyl PG merely has to prevent the protected acid from competing with the free acid for the coupling reagent and esters of various types are usually sufficient, the exact choice depending on the need to deprotect the carboxyl terminus selectively in the presence of the nitrogen PGs and any other reactive functionality. For example, benzyl esters may be cleaved hydrolytically or by catalytic hydrogenolysis; *tert*-butyl esters are cleaved by protic acid (e.g. TFA) and have the advantage that they block nucleophilic attack at the carboxyl carbon to the extent that α-amino *tert*-butyl esters are stable even as their freebases (others may form diketopiperazines); phenyl esters are cleaved rapidly by *catalytic* H_2O_2 in aqueous DMF at pH ≈ 10.5.

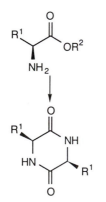

Mechanism of 'OBO' protection (trioxabicyclo[2.2.2]octane)

Diketopiperazine formation (shown) is disfavoured when, e.g., R^2 = *t*-Bu.

CF_3CO_2H trifluoroacetic acid (TFA)

Me_2NCHO dimethylformamide (DMF)

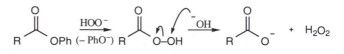

Alkenes (R₂C=CR₂)

Alkenes are reasonable nucleophiles (unless conjugated with electron-withdrawing groups, see OCP #17) that react with a wide range of electrophiles to give addition products. This property may interfere with the smooth course of a desired transformation but may also be used to advantage if the alkene can be regenerated from the adduct at a later stage. Thus 1,2-adducts form by far the widest class of alkene PGs and reagents have been developed to deprotect most of the adducts that can be envisaged, e.g.

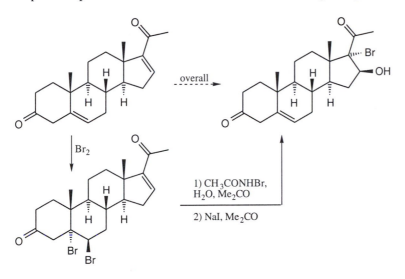

Bromohydrin formation at the less electron-rich alkene can be achieved by protecting the more nucleophilic alkene as a dibromide; NaI is used for the deprotection to avoid bringing the alkene into conjugation with the ketone.

Epoxides (formed by peroxy acid oxidation of alkenes) can be deoxygenated with Ph₃P but this results in inversion of the alkene stereochemistry. Conversely, rhodium carbenoid reagents, generated *in situ* from diazomalonate and a Rh(II) catalyst, remove oxygen from epoxides under mild conditions that retain the alkene stereochemistry.

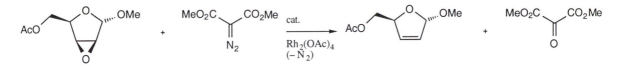

Transition metal complexation to alkenes removes π-electron density from the double bond and creates a sterically hindered environment that leaves the alkene less reactive towards addition reactions. In the example, ligand exchange occurs at the less hindered of the two alkenes (isobutene is liberated), hydrogenation of the ring alkene can then proceed unambiguously, and NaI acts as a reducing agent to decomplex the remaining alkene.

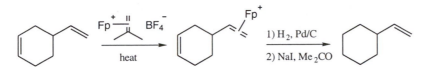

Protection of alkenes as their Diels–Alder adducts (with, for example, cyclopentadiene) is only feasible if the alkene is sufficiently electron deficient to act as a dienophile at reasonable temperatures and this method is usually reserved for conjugated alkenes such as α,β-unsaturated ketones. Reversing the process normally requires high temperatures (often >500°C) but flash vacuum pyrolysis—in which the compound is sucked as an aerosol through a pre-heated tube in the space of a few seconds—is a practical solution.

Consider the consequences of *not* protecting the double bond in the starting ketone in the scheme below.

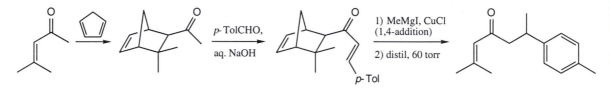

1,3-Dienes

Just as alkenes may be protected as Diels–Alder adducts with cyclopentadiene so 1,3-dienes may be protected as cycloadducts with reactive species such as SO_2 or 4-phenyl-1,2,4-triazoline-3,5-dione. In the following example diene protection is necessary because direct acylation of the imine with the dienyl acid chloride proceeds in very poor yield.

4-phenyl-1,2,4-triazoline-3,5-dione

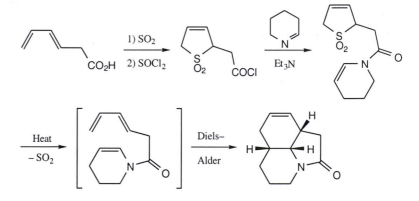

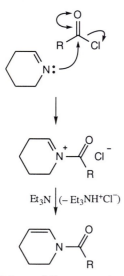

Imine acylation → enamine.

Organotransition metal complexes are also useful in diene protection and the Further reading list should be consulted for a discussion.

Alkynes (RC≡CR)

The mobility of the π-electrons in the carbon–carbon triple bond may lead to undesired reactions with electrophiles and, in general, alkynes are more reactive than alkenes towards hydroboration, acid-catalysed hydration, and redox processes. As a result of this reactivity difference, selective protection of the triple bond in ene-ynes is often necessary and, fortunately, feasible. The roughly tetrahedral complexes formed when alkynes are treated with $Co_2(CO)_8$ are stable to many electrophiles but can be decomplexed easily by oxidation.

The example illustrates overall hydrogenation of an alkene in the presence of an alkyne which requires that the alkyne be protected. Because the cobalt complex poisons hydrogenation catalysts, the overall addition of hydrogen has to be carried out by diimide reduction or, as shown, by hydroboration and protonation:

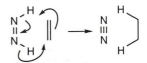

Diimide reduction of alkenes.

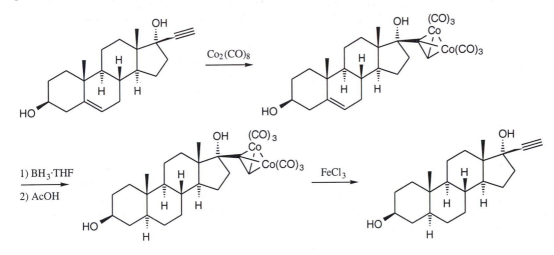

Where the acidity of the alkynyl proton in terminal alkynes presents a problem, temporary replacement of this proton for a base-stable surrogate is required. Silylation has the advantage that there is a wide range of available trialkylsilyl groups that differ in their ease of removal. The C≡C bond in silylated alkynes is resistant to attack by most nucleophiles but will still react with electrophiles (OCP #1) and radicals (OCP #8). Removal of the silyl PG may be achieved with F^- ion (*cf.* hydroxyl protection) or, under milder conditions, with $AgNO_3$ followed by KCN as in the example, which summarises part of a synthesis of an insect hormone:

$$RC\equiv C-H \qquad pK_a \approx 25$$

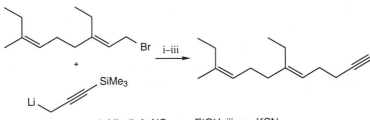

i. Mix; ii. $AgNO_3$, aq. EtOH; iii. aq. KCN

Deprotection pathway:

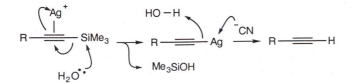

$$2\ Me_3Si-OH$$

$$\downarrow -H_2O$$

$$Me_3Si-O-SiMe_3$$

Trimethylsilanol self-condenses to form the stable siloxane.

1.4 Protecting group devices

When choosing a PG, considerations of what the PG needs to achieve must be balanced against both PG stability and the compatibility of the molecular functionality with the deprotection step. This seems to be a complex decision because of the apparently endless choice of PGs; however, the task is made easier because just a few structural units are used to form the vast majority of PGs in routine use and these structural units define broad areas of cleavage conditions.

We shall refer to these small structural units as *devices* that, alone or in combination, form weak-points (or Achilles' heels) that allow reliable PG removal under specific conditions. Usually, the presence of a particular device in a range of PGs will impart similar deprotection requirements. Identification of a device within a given PG allows us to predict whether the PG will be labile (or stable) under specified reaction conditions. The chart below lists the most important PG devices with two or three examples of their occurrence.

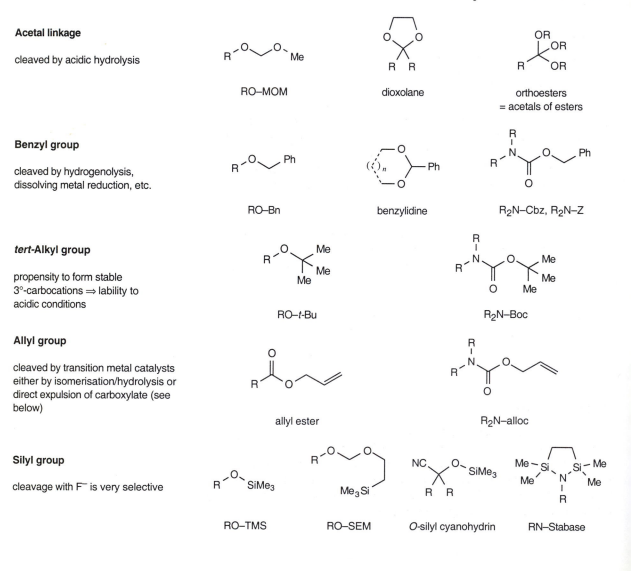

Acetal linkage

cleaved by acidic hydrolysis

RO–MOM   dioxolane   orthoesters
= acetals of esters

Benzyl group

cleaved by hydrogenolysis,
dissolving metal reduction, etc.

RO–Bn   benzylidine   R₂N–Cbz, R₂N–Z

tert-Alkyl group

propensity to form stable
3°-carbocations ⇒ lability to
acidic conditions

RO–t-Bu   R₂N–Boc

Allyl group

cleaved by transition metal catalysts
either by isomerisation/hydrolysis or
direct expulsion of carboxylate (see
below)

allyl ester   R₂N–alloc

Silyl group

cleavage with F⁻ is very selective

RO–TMS   RO–SEM   O-silyl cyanohydrin   RN–Stabase

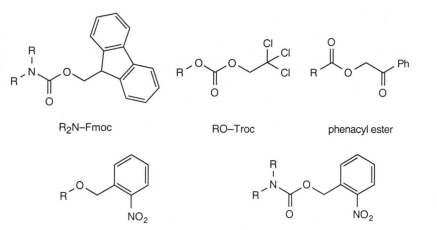

Precursors for β-elimination

(i) base (sometimes after activation)
(ii) e.g. Zn
(iii) F⁻ ion (SEM above)

R₂N–Fmoc   RO–Troc   phenacyl ester

o-Nitrobenzyl group

photolytic cleavage

RO–ONB   R₂N–CO₂–ONB

Decarboxylation

The various PG devices may be attached directly to the protected FG or be separated from it by an intervening $-CO_2-$ unit. The presence of the $-CO_2-$ unit imparts three benefits:

(1)   The PG is introduced by *acylation* (faster) instead of *alkylation*.
(2)   The protected atom is shielded from attack by electrophiles by virtue of conjugation with the carbonyl group.
(3)   Carboxylate is a much better leaving group than RO^- or R_2N^- and deprotection by elimination or substitution processes is much easier.

Following on from point (3), a carboxylate group directly bound to a heteroatom (usually O or N) can collapse by decarboxylation to give the deprotected FG directly (as a salt). Alternatively, loss of CO_2 may take place *after* protonation either during the reaction itself or in the course of aqueous work-up.

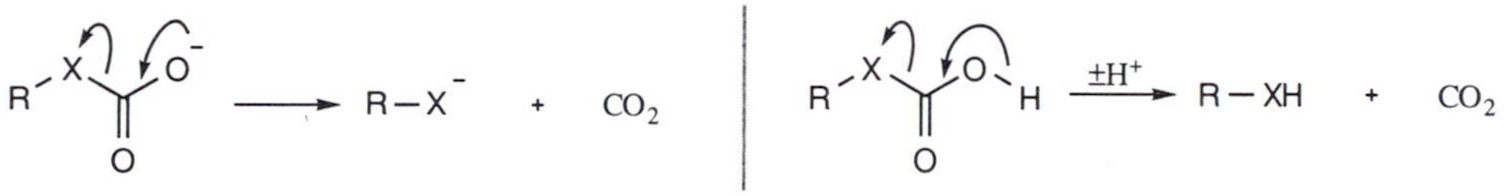

$$R-X\overset{}{\underset{\parallel}{C}}O^- \longrightarrow R-X^- \ + \ CO_2 \qquad \qquad R-X\overset{}{\underset{\parallel}{C}}O\diagdown H \xrightarrow{\pm H^+} R-XH \ + \ CO_2$$

1.5 Temporary protection

If protection is only needed for a single step, it may be possible to employ a *temporary* PG wherein the protected form of the molecule is never actually isolated; instead the PG is introduced as part of the experimental procedure, the desired transformation carried out, and the PG removed at the end of the experiment, normally during the work-up stage. This approach detracts very little from the overall efficiency of a synthesis.

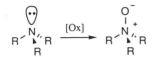

N-oxides are readily formed from amines by oxidation.

–SiPh$_2$*t*-Bu   *tert*-butyldiphenylsilyl
(TBDPS)

meta-chloroperoxybenzoic acid
(MCPBA)

dimethyl dioxirane
(DMDO)

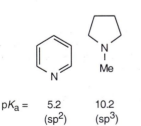

pK$_a$1 = 7.84
pK$_a$2 = 3.04

pK$_a$ =   5.2   10.2
          (sp^2)   (sp^3)

*(pK$_a$ values for amines are given for
the corresponding protonated forms)*

Amines

It is usually impossible to effect peracid epoxidation of an alkene within a molecule that also contains an unprotected amine because most alkyl amines are more readily oxidised than are alkenes (to give the corresponding *N*-oxide leaving the alkene intact). Temporary protection, by protonation (first example below) or by complexation with a Lewis acid (second example) prior to adding the oxidising agent, offers a solution.

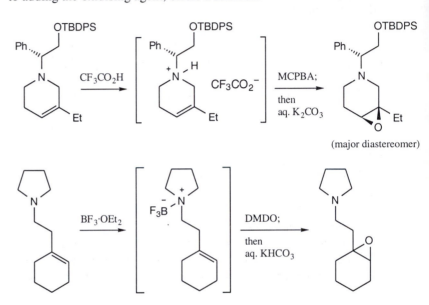

(major diastereomer)

Similarly, methylation of nicotine with MeI is non-selective, a 1:1 mixture of the products of alkylation on the pyridine and pyrrolidine nitrogen atoms being obtained. However, since the pyrrolidine nitrogen is significantly more basic than the pyridine nitrogen, salt formation with HI can provide temporary protection of that nitrogen atom enabling methylation to take place solely on the pyridine nitrogen.

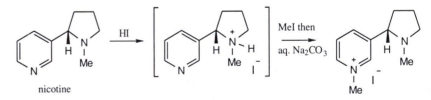

nicotine

Carbonyl groups

In molecules that possess two or more carbonyl groups it is usually possible to find reagents that add selectively to the more electrophilic one. Achieving selective additions to the less electrophilic carbonyl in, for example, a keto-aldehyde is more challenging. A nucleophile that adds to the (more electrophilic) aldehyde to form a temporary adduct prevents further attack at the aldehyde leaving the ketone free to react with other nucleophiles. If the first nucleophile is chosen correctly, the aldehyde adduct may be decomposed at the end of the reaction to reveal the product of overall addition to the ketone.

An example of this strategy employs Me$_2$S as the nucleophile in the presence of a Lewis acid (TMSOTf) to aid its addition to the aldehyde; this leaves the ketone free to be converted into an acetal (using the Noyori conditions, page 32); at the end of the reaction the aldehyde is liberated again by an alkaline work-up.

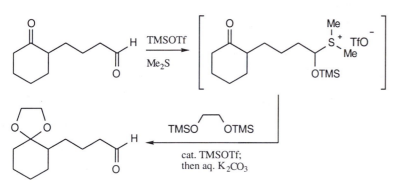

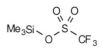

trimethylsilyl trifluoromethane-
sulfonate (TMSOTf)

A similar idea has been used to achieve overall addition to the less reactive ketone in a diketone. In the example below, Ti(NMe$_2$)$_4$ forms a temporary adduct with the more reactive cyclohexadiene carbonyl group, enolate addition to the cyclopentanone ketone follows, and an acidic work-up regenerates the ketone.

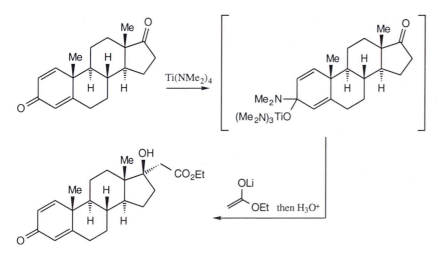

In steroids and other rigid cyclic molecules axial methyl substituents can effectively shield one face of the molecule; in this example addition occurs selectively on the α-face of the carbonyl group.

Lithium amides (R$_2$N–Li) also form temporary adducts with carbonyl groups to provide protection against reaction with organometallic reagents. At the end of the reaction a mildly acidic aqueous work-up decomposes the hemiaminal intermediate to regenerate the carbonyl group. A key step in a synthesis of calicheamicinone—the core structure of calicheamicin, an antibiotic—required addition of an alkynyl lithium reagent to a ketone in the

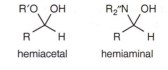

hemiacetal   hemiaminal

presence of an aldehyde which was achieved by temporary protection of the latter with lithium *N*-methylanilide.

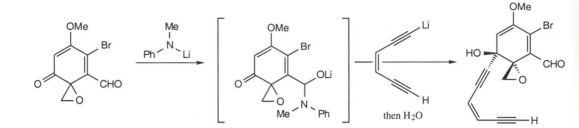

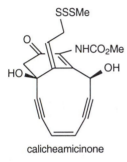

calicheamicinone

This idea derives from studies on the preparation and alkylation of aryllithium reagents in molecules possessing an aldehyde function. In the example temporary protection of the aldehyde (with lithium morpholinide) allows lithium–bromine exchange (with BuLi) and alkylation of the intermediate aryllithium to take place in a single synthetic operation.

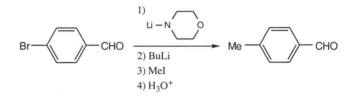

Hydroxyl groups in alcohols and carboxylic acids

Carboxylic acids are reduced rapidly by borane and its derivatives in a synthetically useful method for converting carboxylic acids into 1°-alcohols. This means that attempts to effect alkene hydroboration in molecules containing a $-CO_2H$ function often give complex mixtures of products. Alkene hydroboration can, however, be achieved in the presence of carboxylic *esters* because these are unable to form the reactive tri(acyloxy)borane $[(RCO_2)_3B]$ intermediate that is implicated in the rapid reduction of carboxylic acids. Temporary protection of the $-CO_2H$ group as a TMS-ester enables successful hydroboration and oxidation of unsaturated acids, the intermediate silyl esters being hydrolysed during the oxidation step (alkaline H_2O_2), e.g.

Reminder: trivalent boron is electron deficient; therefore, reactions of boranes are initiated by addition of a nucleophile or complexation to a Lewis base.

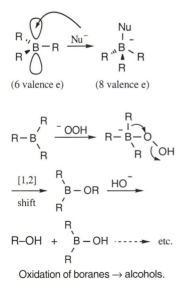

Oxidation of boranes → alcohols.

Temporary silylation was used to protect both a carboxylic acid and a phenolic hydroxyl group which, if left unprotected, interfered with N-sulfonation of tyrosine during a synthesis of MK-383, an antithrombotic drug.

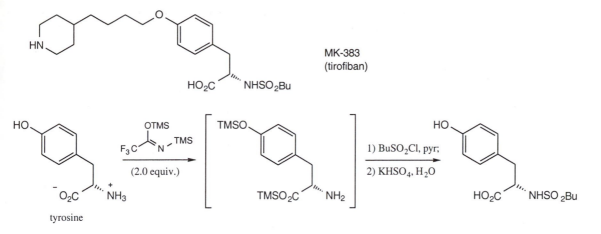

MK-383
(tirofiban)

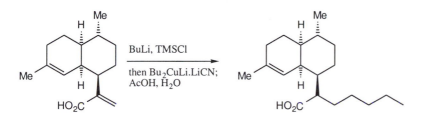

tyrosine

Finally, in a synthesis of analogues of the antimalarial compound artemisinin, temporary protection of an α,β-unsaturated carboxylic acid group as its silyl ester allowed conjugate addition of various cuprates without competing protonation of the organometallic reagent.

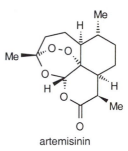

artemisinin

Further reading

Books: J. F. W. McOmie (ed.), *Protective Groups in Organic Chemistry,* Plenum Press, London, 1973; H. Kunz and H. Waldmann, In *Comprehensive Organic Synthesis,* B. M. Trost and I. Fleming (eds.), Pergamon, Oxford, 1991, Vol. 6, p. 631; P. J. Kocienski, *Protecting Groups,* Thieme, Stuttgart, 1994; T. W. Greene and P. G. M. Wuts, *Protective Groups in Organic Synthesis,* 3rd edn., Wiley, New York, 1999; J. R. Hanson, *Protecting Groups in Organic Synthesis,* Sheffield Academic Press, Sheffield, 1999.

Journal articles: M. Schelhaas and H. Waldmann, *Angew. Chem., Int. Ed. Engl.,* **1996,** *35,* 2057; K. Jarowicki and P. Kocienski, *Contemp. Org. Synth.,* **1996,** *3,* 397 & **1997,** *4,* 454; *idem, J. Chem. Soc., Perkin Trans. 1,* **1998,** 4005 & **1999,** 1589.

Cited Oxford Chemistry Primers: #1: S. E. Thomas, *Organic Synthesis. The Roles of Boron and Silicon;* #7: J. Jones, *Amino Acid and Peptide Synthesis;* #8: C. J. Moody and G. H. Whitham, *Reactive Intermediates;* #33: G. H. Whitham, *Organosulfur Chemistry;* #35: G. D. Meakins, *Functional Groups: Characteristics and Interconversions.*

Organometallic PGs: S. G. Davies, *Organotransition Metal Chemistry: Applications to Organic Synthesis,* Pergamon, 1982, Chapter 3.

2 Acid-labile protecting groups

2.1 Introduction

In the protected forms of potentially nucleophilic FGs—alcohols, amines, thiols—the original heteroatom is usually present and capable of reacting with electrophiles unless steric encumbrance or conjugation is completely effective (which it rarely is). Whether or not this leads to deprotection depends on the groups attached to the heteroatom. Reversible protonation or activation with a Lewis acid polarises the electrons in the attached bonds leading to electron deficiency at adjacent centres. If one of the attached groups is capable of supporting a full positive charge then rapid cleavage is likely to follow. This behaviour forms the basis of a large range of *acid-labile* PGs that cleave by forming a stabilised cationic intermediate on addition of either a protic or Lewis acidic reagent (pathway *a* below).

If the attached groups do not readily support a full positive charge nothing further is likely to happen unless a good nucleophile is present (e.g. halide ion) whereupon deprotection may follow by displacement of the activated heteroatom (pathway *b*). Typical Lewis acid–nucleophile combinations include $ZnBr_2$, $TiCl_4$ (with or without added LiI), R_2BBr, TMSI, BBr_3/NaI, $Ac_2O/FeCl_3$, and $BF_3 \cdot OEt_2/Et_4N^+I^-$. The two cases correspond respectively to S_N1 and S_N2 substitution at the PG.

Acidic cleavage of a PG from an FG (X = O, NH, NR, S, CO_2, etc.) with an electrophile (El^+ = H^+ or a Lewis acid). Pathway *a* applies to PGs capable of supporting a full positive charge, pathway *b* applies to nucleophile-mediated cleavage of PGs after Lewis acid activation.

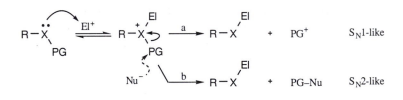

2.2 S_N1-like deprotection

Cation stabilisation

Carbocations are also stabilised 'externally' by polar solvents.

Acid-labile PGs must contain molecular features that stabilise cations. Stabilisation can be achieved by inductive (σ-framework), mesomeric (π-framework), or hyperconjugative (σ–π overlap, OCP #36, page 25) donation of electrons, or by the release of steric strain when the atom bearing the positive charge rehybridises to accommodate it. These features are summarised overleaf for a charge nominally localised on a carbon atom:

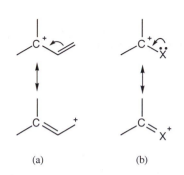

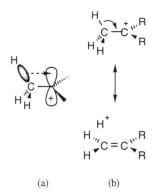

(a)   (b)

(a)   (b)

Inductive

Mesomeric
(a) by adjacent unsaturation
(b) by an adjacent heteroatom (O, N, S)

Hyperconjugative
(a) orbital representation
(b) valence bond picture

tert-Butyl protecting groups

A combination of inductive stabilisation and strain release on ionisation underlie the acid lability of PGs based on the *tert*-butyl group. The -OH group in alcohols and carboxylic acids and the -SH group in thiols can be directly protected by isobutene (condensed or in concentrated solution) in the presence of a protic or Lewis acid (usually H_2SO_4 or $BF_3 \cdot OEt_2$ respectively). Handling isobutene as the gas or in condensed form can be inconvenient; therefore, methods have been devised for generating t-Bu$^+$ from more convenient precursors (e.g. t-BuOH + c.H_2SO_4/$MgSO_4$); t-Bu esters may also be produced by traditional esterification techniques (see page 56). The t-Bu group provides excellent steric shielding of alcohols, impedes nucleophilic attack at the carbonyl group of t-Bu esters, prevents oxidation of thiols to disulfides and, being easily removed by acid, is a popular choice even for the protection of highly functionalised molecules.

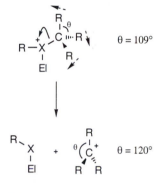

$\theta = 109°$

$\theta = 120°$

Release of steric interactions between large R groups as the carbon rehybridises (sp$^3 \rightarrow$ sp^2).

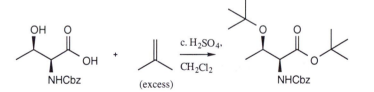

Simultaneous protection of hydroxyl and carboxyl groups.

Cbz = PhCH$_2$OC(=O)– (page 89)

Deprotection takes place rapidly with CF_3CO_2H (TFA), p-TsOH, HCl, and HBr but HF, HCO_2H, and AcOH are more selective. In each case protonation and ionisation are followed by proton loss giving isobutene in an E_1 process, the reverse reaction being minimised by running the deprotections in dilute solution.

	pK_a
HBr	–9
HCl	–7
p-TsOH	–6.5
TFA	0.23
HF	3.2
HCO_2H	3.8
AcOH	4.8

Protection

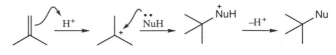

Protection and deprotection proceed through the same intermediates and are mechanistic opposites.

NuH = ROH, RCO$_2$H, RSH

Deprotection

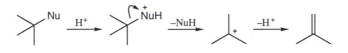

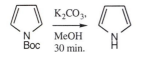

If the nitrogen forms part of an aromatic ring, Boc cleavage can be rapid (the N lone pair contributes to the aromatic π-system and cannot stabilise the carbamate carbonyl to the same degree).

   Amines can also be protected by the *t*-Bu group but the acidic conditions required for direct *tert*-butylation preclude its use, the amine protonating in preference (R$_2$NH → R$_2$NH$_2^+$, the ammonium cation is non-nucleophilic, *cf.* page 5). Instead, linking the nitrogen atom and the *t*-Bu group by a carboxyl group (page 13) allows easy introduction of this *tert*-butyloxy-carbonyl (Boc) PG by acylation. *t*-Butyl chloroformate (*t*-BuOCOCl) is unstable and the anhydride [(*t*-BuOCO)$_2$O, Boc$_2$O] is generally the reagent of choice; Boc-ON in the presence of an amine base is also widely used. Boc protection removes most of the nucleophilicity of the amine nitrogen and the PG itself is stable to hydrogenation, bases, and most nucleophiles.

Mechanistic parallels in the Boc protection of amines with Boc$_2$O and Boc-ON. The bases shown are required in the protection of amino acids or other amines where the nitrogen is protonated in solution.

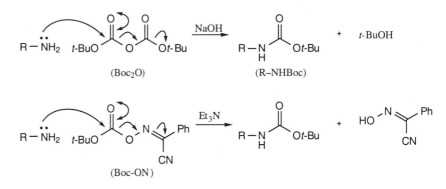

   During deprotection, protonation on the carbonyl oxygen and loss of *t*-Bu$^+$ result in a carbamic acid which decarboxylates rapidly to give the amine.

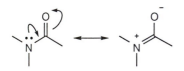

Conjugation increases electron density on the amide oxygen.

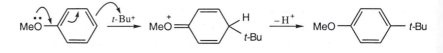

   TFA is the acid of choice (fast acting, volatile, and easily removed by evaporation) but the liberated *t*-Bu cation is a powerful electrophile and, where this leads to side-reactions, a Lewis acid is used in the presence of a scavenger, such as anisole or thioanisole, to trap it preferentially.

Friedel–Crafts alkylation of anisole and other scavengers reduces side-reactions induced by the *t*-Bu cation.

N-Boc derivatives are more easily cleaved than are *t*-Bu ethers and *t*-Bu esters, and *S*-Boc derivatives can be left unscathed if mild (room temperature) conditions are used. This may reflect selective protonation of the *N*-Boc carbonyl group which is expected to be the most basic, considering the more favourable conjugation between the nitrogen lone pair and the carbonyl π-system.

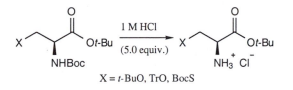

X = *t*-BuO, TrO, BocS

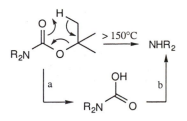

Thermolytic Boc-deprotection is useful if acidic conditions are not tolerated (step *a* liberates isobutene; CO_2 is lost in step *b*).

Ph₃C-   trityl-   (Tr-)

Benzylic protecting groups

Stabilisation of positive charge by conjugation with an alkenyl π-system also provides a basis for acid lability. A lone unsubstituted double bond (i.e. an allyl PG) does not confer enough stability to a cation for ionisation to occur under normal protic conditions but if the double bond is substituted appropriately then cleavage can be significant, at least in the case of carboxyl protection, where the deprotection of methallyl esters follows an $A_{Al}1$ pathway. Under comparable conditions benzyl esters are only partially cleaved which suggests a relative order of cation stability: allyl < benzyl < methallyl.

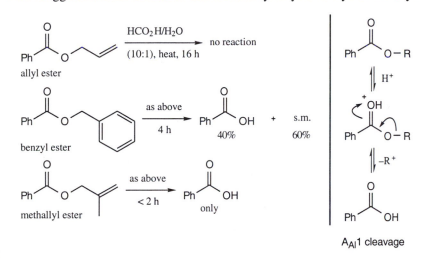

Only a few of the many combinations of FGs and PGs based on benzyl functionality are routinely cleaved by acidic hydrolysis; benzyl PGs are more often chosen because of their relative ease of cleavage by redox processes. In this section we will concentrate on those benzyl PGs whose acid lability is valued in synthesis.

A single unsubstituted benzyl group can rarely be cleaved by protic acids alone and Lewis acid–nucleophile combinations are more reliable (Section 2.3). However, two or more phenyl rings (or one if it is substituted with an electron-donating group, e.g. OMe) *are* sufficient to ensure ionisation, and PG cleavage can be relied upon after addition of the usual proton sources.

For *alcohol* protection the Ph₃C- (triphenylmethyl, trityl, Tr) group, historically developed for carbohydrate chemistry and later aiding the construction of oligonucleotides, is the most important. Only 1°-alcohols react at a useful rate with trityl chloride in pyridine (S_N1 mechanism) so selective protection of these hydroxyls is possible (1); this feature is a major reason for using this PG. In the presence of the more powerful base 1,8-diazabicyclo[5.4.0]undec-7-ene (DBU), even 2°-alcohols can be tritylated (2).

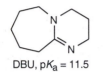

DBU, pK_a = 11.5

(pK$_a$ values for amines are given for the corresponding protonated forms)

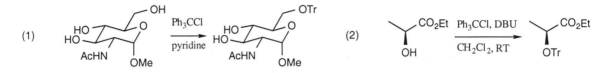

(1)

(2)

In addition to providing selective protection of 1°-hydroxyl groups the trityl group offers good resistance to bases but is easily removed by acid treatment, TFA, HCO₂H, and AcOH offering progressively milder cleavage conditions (pK_as: page 19). Formic acid in ether is sufficiently selective to cleave trityl ethers in the presence of other acid-sensitive groups including acetals (page 24) and TBS ethers (page 77).

t-BuMe₂Si- *tert*-butyldimethylsilyl (TBS)

PhC(=O)- benzoyl (Bz)

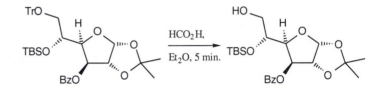

The rate of trityl group cleavage can be fine-tuned by substituents on one or more of the phenyl rings; electron-donating groups favour cation formation which speeds the cleavage process, and vice versa. This is particularly useful in the protection of the 5′-OH in nucleosides where more acid-labile variants of the trityl PG can be removed without competing loss of the base (which can be a problem with the trityl group itself). Below is a comparison of the times required to effect complete hydrolysis of trityl (Tr), *p*-methoxyphenyldiphenylmethyl (MMTr), di(*p*-methoxyphenyl)phenylmethyl (DMTr), and tri(*p*-methoxyphenyl)methyl (TMTr) ethers.

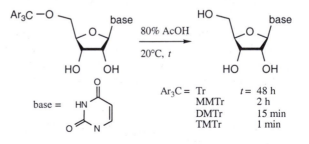

Ar₃C =		*t* =
Tr		48 h
MMTr		2 h
DMTr		15 min
TMTr		1 min

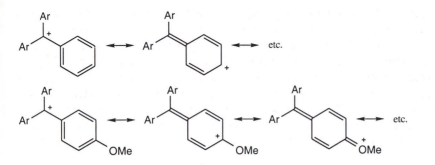

The need for *amine* protection during amino acid and peptide synthesis has given rise to a variety of benzyl PGs that enjoy widespread use rather than being of merely academic interest.

For example, whilst simple benzyl amines are not efficiently cleaved under protic conditions N-trityl amines are. These are readily prepared with either TrCl or TrBr + Et₃N and, once on, are very stable to basic reagents. However, N-tritylation of α-amino acid derivatives sterically encumbers the carboxyl group inhibiting ester hydrolysis and peptide formation which limits its applications. A second drawback is that Tr amines can be *so* labile to acids that handling and purifying them is impractical; however, if two of the phenyl rings are joined by a single bond, the ease of cation formation is dramatically reduced.

This modified Tr PG—the 9-phenylfluorenyl (PhFl) group—shares many features in common with Tr (stability to bases and organometallic reagents, steric bulk and ability to shield adjacent carboxyl functionality, solubility in organic solvents) but is around 6000 times more stable to acid. The fluorenyl cation is a 12π-electron system, is therefore antiaromatic, and its formation is comparatively unfavourable.

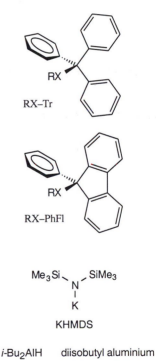

RX
RX–Tr

RX
RX–PhFl

Me₃Si—N—SiMe₃
|
K
KHMDS

i-Bu₂AlH   diisobutyl aluminium hydride   (DIBAL)

Fluorenyl cation (12 π electrons)   Cyclopentadienyl cation (4 π electrons).

Introduction of the PhFl group can be achieved under conditions similar to those used for trityl protection (e.g. PhFlBr, Et₃N) but deprotection requires more strongly acidic conditions (aq. TFA, CH₃CN) (hydrogenolysis or dissolving metal reduction is also effective, Section 5.3). In the synthesis of unnatural amino acids the steric bulk of this PG disfavours deprotonation at the α-position of aspartic acid derivatives and prevents nucleophilic addition to the adjacent carboxyl group; DIBAL reduces the γ-carboxyl group even though that ester is also quite hindered (by an adjacent 4°-centre).

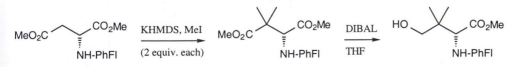

MeO₂C⌒⌒CO₂Me   KHMDS, MeI   MeO₂C⌒⌐CO₂Me   DIBAL   HO⌒⌐CO₂Me
| (2 equiv. each) | THF |
NH-PhFl   NH-PhFl   NH-PhFl

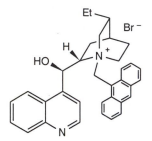

The chiral catalyst (cat.* in the scheme below).

Diphenylmethylene imine derivatives represent a further acid-labile benzyl PG for amines. Their most useful application is in the synthesis of amino acids by alkylation of protected glycine enolates either racemically or enantioselectively with a homochiral phase transfer catalyst derived from cinchonidine, an isoquinoline alkaloid. Protection in this way removes problems associated with the presence of a free amino function and the extra double bond facilitates deprotonation by adding stability to the enolate. When no longer required, this PG can be removed by mild acidic hydrolysis; citric acid is often used because its high solubility in water makes its removal by extraction a trivial process.

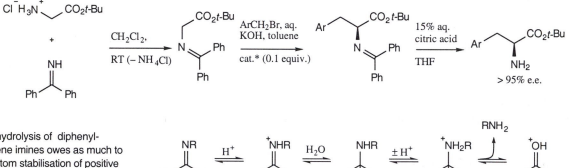

Acidic hydrolysis of diphenyl-methylene imines owes as much to heteroatom stabilisation of positive charges as it does to delocalisation over the phenyl rings.

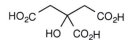

citric acid, pK_a = 3.1

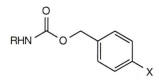

X = H: RNHCbz (a.k.a. RNHZ)
X = MeO: RNHMoz.

Carbamates dominate nitrogen PG chemistry, combining ease of introduction and effective protection of the amine towards electrophiles with versatility because the cleavage conditions can be specified by appropriate choice of the *O*-alkyl substituent. In the benzyloxycarbonyl (Cbz or Z) group the single phenyl ring is insufficient to confer significant protic acid lability unless a good nucleophile is present (*cf.* Section 2.3); usually, strongly acidic conditions are needed (e.g. HBr, AcOH) and TFA cleaves this group significantly more slowly than the Boc group (⇒ selective deprotection). However, the *para*-methoxybenzyl (Moz) analogue is, as you should expect, more labile and can be cleaved with *p*-TsOH in acetone/CH_3CN or 10% TFA in CH_2Cl_2; in fact, this PG is *more* reactive under these conditions than Boc so a useful series is established graded for ease of removal by protic acid: RNHMoz > RNHBoc > RNHCbz. Many variants are possible and analogous *carbonates* have been used for alcohol protection.

Acetal protecting groups

From this discussion it should be evident that heteroatoms—particularly oxygen—can tip the balance and convert a relatively poorly acid-labile PG (e.g. benzyl) into one (e.g. *para*-methoxybenzyl) that may be cleaved reliably with mildly acidic reagents. Because a single phenyl ring exerts a relatively weak cation-stabilising influence, it can be dispensed with altogether if an oxygen atom is present on what was the benzylic position. This forms the basis of a second set of acid-labile PGs, those based on acetals (and aminoacetals, dithioacetals etc.):

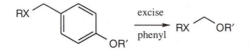

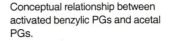

Conceptual relationship between activated benzylic PGs and acetal PGs.

As a reminder, an acetal is the product obtained when an aldehyde or ketone reacts with two molecules of an alcohol with liberation of water; a new compound is formed wherein the carbonyl oxygen is replaced by two alkoxy groups (if a diol is used a cyclic acetal results). The process is catalysed by both protic and Lewis acids and, being reversible, requires removal of water to drive the reaction towards the acetal. To effect hydrolysis water is added, or acetal exchange with a large excess of an expendable carbonyl compound—usually acetone—can be used. In general:

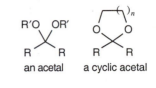

an acetal   a cyclic acetal

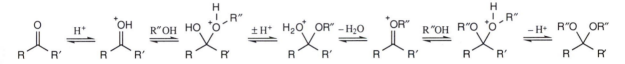

Since both an alcohol and a carbonyl component combine to form a molecule which is readily decomposed by acidic hydrolysis, this represents a unified PG scheme for both these FGs.

Acetal protecting groups for alcohols

Protecting an alcohol as a symmetrical acetal by direct condensation with a carbonyl compound is rarely done because synthetic intermediates are often complex structures that are too bulky to be linked by a single carbon atom. If there is more than one OH group present a large number of acetals could be produced, and a chiral but racemic alcohol will lead to a series of diastereomeric acetals (*cf.* page 27), which makes identification and purification complicated. Overall, it is much more efficient to prepare *mixed* acetals where the second alkoxy group is expendable. By starting with an enol ether, which yields an oxonium ion after protonation, or a chloromethyl ether, which can lose Cl⁻ ion as the first step of an overall S_N1 substitution, mixed acetal products can be produced with predictable constitution.

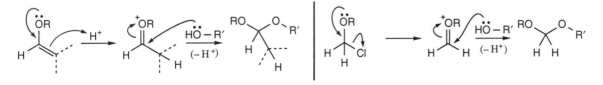

These two classes of reagents differ in the conditions needed to introduce them. Enol ether PG reagents require acid catalysis—a proton is added and removed for each product-forming reaction—but acetal formation with a chloromethyl ether liberates HCl which must be removed if unwanted reactions are to be avoided; a 3°-amine (e.g. Hünig's base, *i*-Pr₂NEt) or NaH is added for this purpose.

The simplest acetal PG is the methoxymethyl (MOM) ether which is easily introduced onto 1°- and 2°-alcohols using chloromethyl methyl ether (MOM-Cl) and Hünig's base; 3°-alcohols need MOM-I (formed *in situ* with

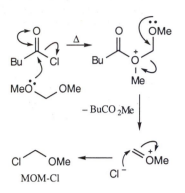

Synthesis of MOM-Cl (safer than
HCHO + MeOH + HCl).

MOM-Cl

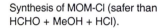

camphorsulfonic acid (CSA)
(cheap, soluble in organic solvents;
the fact that it is chiral is unimportant).

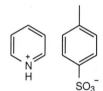

pyridinium *p*-toluenesulfonate (PPTS)

p-TsOH   pK_a = –6.5
pyridine·H$^+$ pK_a = 5.2

RO–MME (20) > RO–EE (*ca.* 10) >
RO–4-MeO-THP (3) > RO–THP (1)

Approximate relative rates of cleavage
under protic hydrolysis conditions.

NaI) suggesting that the protection mechanism is at the S_N1–S_N2 borderline. The high toxicity of MOM-Cl has led to the use of alternative reagents (e.g. $(MeO)_2CH_2$ + $FeCl_3$) that work on the same principle.

Aside from its ease of introduction, the principle virtues of the MOM group are that it is small and robust. MOM-protected alcohols provide little steric impediment to subsequent reactions and they can survive a wide range of reaction conditions. Sometimes they are *so* stable that cleaving them again can be a problem and the protic conditions needed to remove them (e.g. conc. HCl, MeOH) are sufficiently harsh that almost all the other acid-labile PGs are also cleaved. Milder reagents, based on Lewis acid/nucleophile combinations, are much more selective (Section 2.3).

Acetal PGs for alcohols formed from enol ethers include tetrahydropyranyl (THP) and ethoxyethyl (EE) derivatives, formed respectively from dihydropyran and ethyl vinyl ether with acid catalysis (by the general mechanism shown above), the acids of choice being CSA, *p*-TsOH, or PPTS (milder).

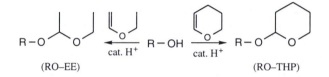

(RO–EE)   (RO–THP)

Both of these groups are very widely used in synthesis, offering similar benefits to MOM protection but being more readily cleaved by protic conditions as a result of the extra alkyl substituent at the acetal carbon which adds stability to the liberated oxonium ion. Even more rapid cleavage ensues when a second alkyl group is present on the acetal carbon; treatment of alcohols with 2-methoxypropene and an acid catalyst produces 1-methoxy-1-methylethyl (MME) ethers which can be cleaved subsequently with aqueous AcOH. Similarly, 4-methoxy-THP ethers can be prepared from 4-methoxy-dihydropyran and then cleaved when desired under acidic conditions.

(RO–MME)

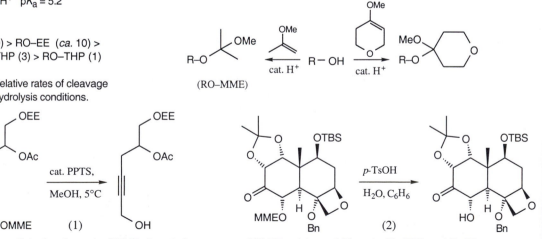

Selective cleavage of MME ethers in the presence of (1) EE and Ac, and (2) acetonide, TBS, and Bn PGs.

A disadvantage of the EE and THP PGs is that they introduce a stereogenic centre at the newly-formed acetal carbon. If the protected alcohol is chiral then two diastereomers will result; if more than one alcohol is protected the total number of diasteromers can become impractical. For example, THP protection of *R*-butane-1,3-diol could result in four diastereomeric products:

Mixtures of diastereomers are difficult to crystallise, purify, and characterise spectroscopically.

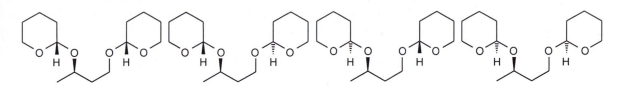

This is another advantage of 4-methoxy-THP and MME derivatives since no new stereogenic centre is introduced with these PGs.

Acetal protecting groups for diols

Direct condensation* with carbonyl compounds *is* useful for the protection of 1,2- and 1,3-diols as cyclic acetals because both kinetic and thermodynamic factors act cooperatively to favour intramolecular acetal formation over the various intermolecular possibilities. The most useful cyclic acetals are those formed between diols and either acetone or benzaldehyde although those based on formaldehyde, acetaldehyde, and 5–7-membered cyclic ketones are also frequently used. They can all be introduced onto the diol with the carbonyl compound itself, or with the synthetic equivalents listed below, plus an acid catalyst [e.g. HCl, HBF_4, H_2SO_4 (which also removes water), or *p*-TsOH]:

*Condensation: a reaction in which water is liberated.

≡ formaldehyde (HCHO)	≡ acetaldehyde (CH$_3$CHO)	≡ acetone (CH$_3$COCH$_3$)	≡ benzaldehyde (PhCHO)

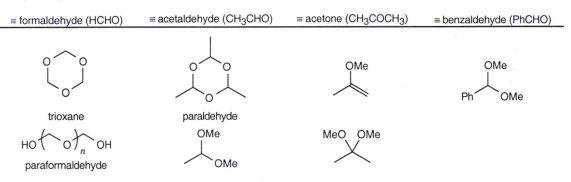

In the presence of an acid catalyst these synthetic equivalents are capable of producing the oxonium ion intermediate necessary for nucleophilic attack by one of the hydroxyls of the diol to be protected. The idea is illustrated for acetonide formation:

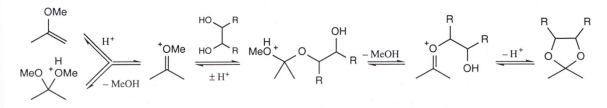

1,3-dioxane

1,3-dioxolane

Since all these derivatives are formed under similar conditions how do we choose which acetal PG to use for a particular situation? A useful generalisation is that, given a choice of reacting with a 1,2- or 1,3-diol, ketones preferentially form five-membered cyclic acetals (i.e. the 1,2-diol is protected) whereas aldehydes form six-membered cyclic acetals as major products (the 1,3-diol is protected). These processes are reversible and therefore are subject to thermodynamic control, the major acetal product being that which is most stable *under the reaction conditions*; the italics are important as concentration, temperature, choice of catalyst, etc. can influence product ratios. Although the dioxane ring is inherently more stable than a comparable dioxolane, six-membered (dioxane) *ketone* acetals cannot avoid destabilising 1,3-diaxial interactions between one of the alkyl groups and *cis*-axial substituents in either chairlike conformation; on the other hand, the single alkyl group in *aldehyde*-derived dioxanes is free to take up an equatorial position to avoid this effect.

Acetonide dioxanes

Benzylidene dioxanes

In five-membered ring acetals (dioxolanes) axial and equatorial sites are less clearly defined and, on average, substituents on the acetal carbon are further away and therefore less able to interact unfavourably with other atoms around the ring. For this reason dioxolanes formed from aldehydes do not usually show a distinct preference for a particular configuration at the acetal carbon.

Distances (Å) between substituents in low-lying conformations of 2,2-dimethyl-dioxane and -dioxolane.

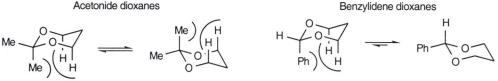

envelope   half-chair = twist

In principle, then, given a polyhydroxylated molecule it should be possible to pair 1,2-diols or 1,3-diols depending on the choice of carbonyl component—aldehyde or ketone—again bearing in mind that the reaction conditions may be critical. In the example of acetonide formation in glucose (page 2), protection is accompanied by formation of the furanose product driven by the preferred formation of dioxolane rings. In contrast, treatment of glucose with benzaldehyde under equilibrating conditions proceeds to form the dioxane product that links the 4,6-diol pair.

Benzylidenation of glucose gives the dioxane product with the phenyl group in an equatorial site (*cf.* page 2).

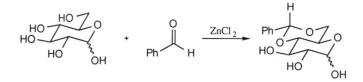

These principles also extend to acyclic cases, e.g.

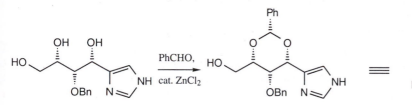

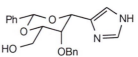

although the results are not always as expected. In the example below, acetal formation takes place between the 1,3-diol pair that leads to an *axial* hydroxyl; this has been ascribed to favourable hydrogen bonding between the proton of this OH and either one or both of the acetal oxygens (pairing the other terminal 1,3-diol would give a product with all groups equatorial).

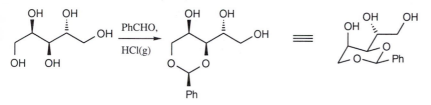

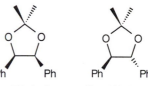

Two 3H singlets, one for each Me group.

One 6H singlet, the Me groups are chemically equivalent.

Formation of cyclic acetals from acyclic diols restricts their conformation and greatly facilitates the solution of stereochemical problems by ^{1}H NMR analysis.

Knowledge of the relative stabilities of the available cyclic acetals allows an appropriate level of protection (against bases, organometallics, etc.) to be selected without making eventual deprotection unnecessarily difficult. Acetal lability to acid parallels the stability ($\Rightarrow$ ease of formation) of the oxonium ion generated when the dioxane/dioxolane first breaks open, i.e.

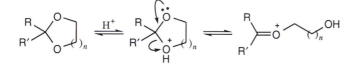

The stability of this oxonium ion is raised by alkyl R and R′ substituents (inductive, hyperconjugative); if one or both of these is an aryl substituent further stabilisation of the positive charge results through conjugation. Furthermore, release of ring strain in the ring-opening process will favour ionisation and the effect of changing the hybridisation at the acetal carbon from sp^3 to sp^2 also needs to be taken into account. These factors combine to give the following general principles concerning ease of removal:

(1) acetonide > ethylidene > methylene (which are difficult to remove);
(2) *trans*-fused dioxolanes > dioxanes > *cis*-fused dioxolanes;
(3) acetals of 1°-alcohols > acetals of 2°-alcohols;
(4) cyclopentylidene ≈ cycloheptylidene > acetonide >> cyclohexylidene;
(5) electron-donating groups on aryl substituents (if present) favour cleavage (e.g. *p*-methoxybenzylidine acetals hydrolyse 10× faster than benzylidene acetals);

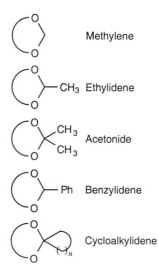

Methylene

—CH$_3$ Ethylidene

CH$_3$
CH$_3$ Acetonide

—Ph Benzylidene

Cycloalkylidene

Cyclic acetal derivatives of diols.

(6) benzylidene dioxanes *can* be hydrolysed in the presence of acetonide dioxolanes but it is usually easier to take advantage of their ability to be removed reductively (Section 5.3).

Selective protection of *cis*-1,2-diol pairings in carbohydrates is readily achievable since *cis*-fused dioxolanes are considerably more stable than their *trans*-fused counterparts (point 2 above). What if we need to protect selectively a *trans*-1,2-diol pairing? An elegant solution to this problem extends the idea of acetal formation to include the reaction between diequatorial 1,2-diols and 1,2-diketones in MeOH to give diacetal products. The formation of these diacetals is subject to thermodynamic control (the kinetics are complex), the major stabilising features being (i) formation of a *trans*-decalin ring system, (ii) placing the maximum number of alkyl groups in equatorial sites, (iii) maximisation of the number of anomeric effects (see OCP #s 36 and 54). Although many important applications of this principle have been described—notably for oligosaccharide synthesis—in its simplest form the chemistry offers a useful complement to carbohydrate protection by acetonide and benzylidene formation. Two examples are given below.

Butanediacetal (BDA) protection of *trans*-(diequatorial)-1,2-diol pairs either by direct condensation with butane-2,3-dione or by acetal exchange.

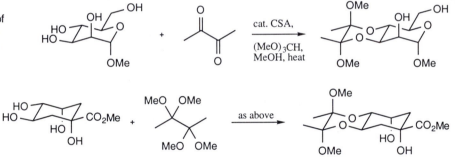

Protection of other difunctional molecules

Other combinations of nucleophilic FGs also participate in forming species analogous to acetals:

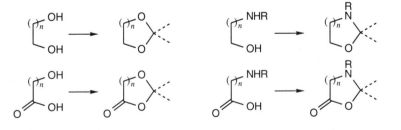

These derivatives are widespread in amino acid synthesis where a combination of amine, alcohol, and carboxylic acid functionality in close proximity necessitates extensive PG schemes. In the example below, condensation of L-proline with pivalaldehyde (*t*-BuCHO) protects the amine and carboxyl functions during enolate alkylation which would otherwise be

impossible. During the protection step (thermodynamic control) the bulky
t-Bu group assumes the less-hindered position on the *exo*-face of the bicyclic
product. In the subsequent alkylation step the electrophile (MeI) approaches
the enolate preferentially on the same less-crowded *exo*-face (kinetic control).
These two features combine to relay the configuration of the original amino
acid to that of the alkylated product.

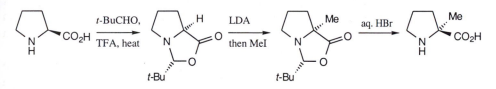

Meldrum's acid, an acetal derivative of a diacid, is formed from acetone and
malonic acid with dehydration assisted by acetylation. This compound is a
usefully protected form of malonic acid that undergoes alkylation and aldol
chemistry to give products that are easily deprotected by mild acid hydrolysis.

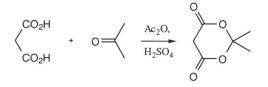

Meldrum's acid: pK_a = 4.83
(π-system automatically
orientated correctly for overlap
but >99.5% diketo tautomer).

Cf. dimedone (pK_a = 5.2),
mainly enolic.

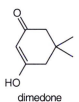

dimedone

Acetal protecting groups for aldehydes and ketones

The discussion on diol protection applies equally to the protection of
carbonyl compounds as acetals (acetals incorporating one of the various PG
devices that allow alternatives to acidic hydrolysis will be discussed
separately). Dimethyl acetal derivatives of carbonyl compounds are obtained
by warming the compound in MeOH and/or trimethylorthoformate
[(MeO)$_3$CH] with an acid catalyst (dry HCl, *p*-TsOH, etc.) although many
variations are possible. The trimethyl orthoformate procedure requires no
added drying agent because the liberated water is consumed by the orthoester
to generate methyl formate.

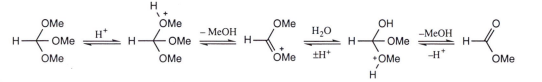

Dimethyl acetals are rather labile and those prepared from ketones are often
too unstable to be isolated; therefore, this PG is usually reserved for aldehyde
protection. When applied to α,β-unsaturated aldehydes in the presence of, for
example, HBr, acetal formation is accompanied by 1,4-addition of the halide
ion yielding useful precursors for further reactions after which the acetal can
be hydrolysed easily without affecting many other acid-labile groups
(including dioxolanes, epoxides, silyl ethers, etc.).

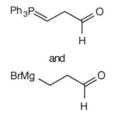

and

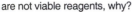

are not viable reagents, why?

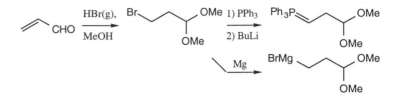

There is an alternative to the usual H⁺/alcohol procedure that is especially useful for the protection of acid-sensitive substrates, when selectivity between two or more carbonyl groups is needed, or when the alcohol component is functionalised. Treatment of aldehydes or ketones with two equivalents of a trimethylsilyl ether (ROTMS, page 76) and a catalytic amount of TMSOTf at low temperature (−78°C is usual) results in clean acetalisation by a process that is mechanistically equivalent to the standard procedure except that a trimethylsilyl group takes the place of a proton in each step.

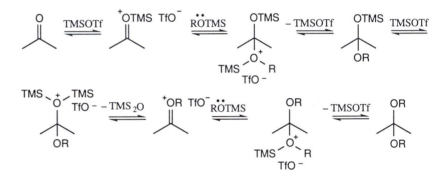

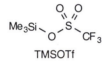

The original carbonyl oxygen ends up in TMS₂O (volatile) rather than H₂O so that a separate dehydrating agent is not required. Use of the bis(trimethylsilyl)ether of a diol affords the corresponding cyclic acetal. An example of this 'Noyori acetalisation procedure' is given on page 15.

Cyclic acetals are more frequently used in synthesis than acyclic acetals as they are less labile and chromatographic purification on (mildly acidic) silica gel is not normally complicated by partial hydrolysis. Dioxolanes, formed from the carbonyl compound plus ethan-1,2-diol under the usual conditions (cat. H⁺, −H₂O), have proven most popular and the Noyori conditions also work well here. The comparison of dioxolane acetals (from 1,2-diols) and dioxane acetals (from 1,3-diols) extends to carbonyl protection; the unavoidable axial alkyl group in ketone dioxanes renders these acetals the most labile to deprotection. The ease of acidic hydrolysis is:

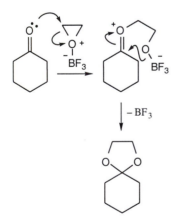

Dioxolanes may also be formed with epoxides and a Lewis acid catalyst.

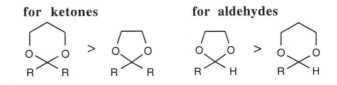

Cyclic acetals are ideal for preventing nucleophilic addition to the carbonyl group and for avoiding potentially problematic deprotonation of acidic α-protons as illustrated very simply by protection of ethyl acetoacetate, its reaction with a Grignard reagent, and subsequent deprotection by hydrolysis (accompanied by dehydration by an E1 mechanism).

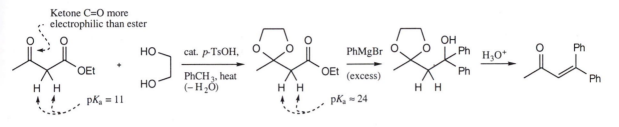

Selectivity between two carbonyl groups is also possible. In general: (1) aldehydes are protected in the presence of ketones; (2) for two aldehydes (or two ketones) protection takes place at the less hindered of the two carbonyls (assuming neither or both are conjugated); (3) if one is conjugated the other is protected kinetically (assuming the steric environments are similar).

C_2-symmetric, chiral 1,2- and 1,3-diols form chiral acetals that are capable of diastereoselective reactions after complexation with Lewis acids at the less-hindered lone pair of electrons on the less hindered oxygen, e.g.

In the example the right-hand acetal oxygen is less hindered on the face that bears the alkenyl part of the cyclohexene.

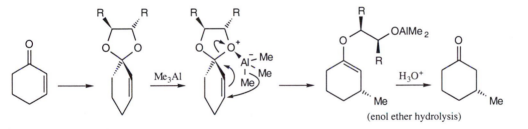

(enol ether hydrolysis)

In the case R = CONMe$_2$ the final product is obtained in 76% e.e.; this illustrates the use of an acetal PG as a chiral auxiliary to direct 1,4-addition from a specific face of cyclohexenone (see OCP #63).

Acetals of esters: orthoesters

The formation of orthoesters [RC(OR′)$_3$] from esters is analogous to acetal formation from aldehydes and ketones but their direct preparation from the acid is usually an inefficient process. Fortunately, alkyl nitriles react with alcohols in a two-stage process to give orthoesters in good yield.

OEt
⊢OEt
OEt
triethylorthoacetate,
an orthoester

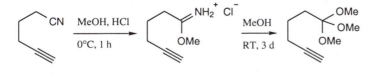

OMe
H⊢OMe
NMe$_2$
N,N-dimethylformamide dimethyl
acetal, an orthoamide

Orthoester derivatives of carboxyl groups provide similar levels of protection as acetals do for aldehydes and ketones except they are somewhat more acid labile, the cation formed after ejection of one alkoxy group being stabilised by the *two* that remain. In fact, the Mg(II) component of Grignard reagents is sufficiently Lewis acidic to induce ionisation and alkylation above room temperature.

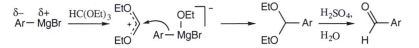

Although this is a useful method for homologating Grignard reagents, in the context of PG chemistry this is something to be avoided; fortunately, at the usual temperatures of organometallic reactions (–78 to 0°C), this process is too slow to compete with additions to reactive electrophiles such as carbonyl groups and alkyl halides. Cuprates and organolithium reagents are insufficiently Lewis acidic to induce this reaction even at elevated temperatures (a Lewis acid, such as $BF_3 \cdot OEt_2$, can be added to activate acetals to react with these organometallics).

Acyclic orthoesters, i.e. those prepared from three equivalents of a monohydric alcohol, can be too labile to be useful in extended synthetic sequences but constrained versions, based on triols, are less fragile and survive column chromatography on silica gel (a factor that often dictates whether or not a particular PG is considered useful). The OBO PG was introduced on page 8 so here we shall just include an example of its use

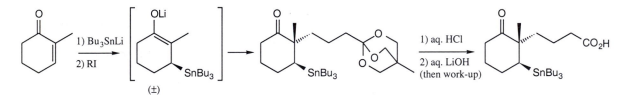

RI = I-(CH$_2$)$_3$-OBO

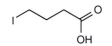

This electrophile cannot be used in enolate alkylation (acidic proton).

and the course of its cleavage which is a two-step process consisting of mild acid hydrolysis followed by saponification of the so-formed ester.

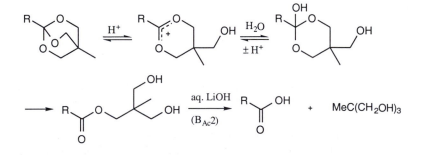

Orthoesters derived from *cis,cis*-1,3,5-trihydroxycyclohexane—trioxaadam-antyl (TOA) derivatives—are less easily cleaved than OBO derivatives. The oxonium ion first formed during OBO deprotection can easily adopt a chairlike conformation to allow (stereoelectronically preferred, see OCP #36) axial attack by water. By contrast, the TOA-derived oxonium ion, which is already in a favourable chair conformation, is hindered by the liberated hydroxyl group towards nucleophilic attack and can only be attacked axially once the second ring has adopted a relatively unfavourable boat conformation.

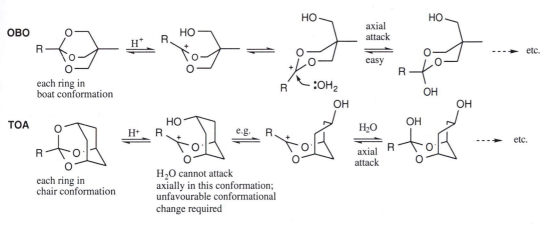

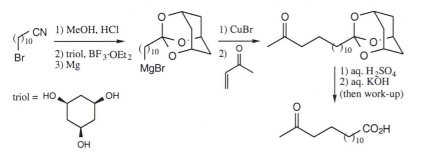

As a result TOA derivatives are entirely compatible with Grignard reagents:

This Grignard reagent would be destroyed by self-protonation.

2.3 Nucleophile-assisted deprotection

In Section 2.2 we discussed protected FGs that, after protonation, ionise essentially spontaneously to liberate the FG and generate a stabilised cation. However, many PGs do not possess sufficient cation-stabilising functionality to induce spontaneous ionisation (benzyl is borderline) and protonation merely enhances the FG's ability to act as a leaving group in an S_N2-like process provided an efficient nucleophile is present. This reactivity forms the basis of another category of acidic cleavage conditions that are based on a combination of a Lewis acid and a nucleophile.

PGs that *require* these cleavage conditions (as opposed to the usual protic conditions) include methyl ethers, certain benzylic PGs, and those containing sulfur (strongly nucleophilic, weakly basic).

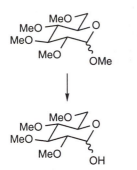

Superficially, this looks like a methyl ether cleavage but it is a partial hydrolysis of an acetal to a hemiacetal (*cf.* page 25). The (normal) methyl ethers remain unscathed.

Methyl ethers

Methyl ether derivatives of alcohols are normally prepared using the Williamson ether synthesis (NaH, MeI, or Me_2SO_4) or, for base-sensitive alcohols, MeI and Ag_2O. Methyl ethers are essentially inert to even strongly acidic or basic conditions with the corollary being that rather forcing conditions are often needed to remove them again. The methyl cation is insufficiently stable to allow cleavage by an S_N1 pathway and alkoxide ion is a poor leaving group (but ArO^- is reasonable); therefore, Lewis acid activation is required to facilitate an S_N2 pathway. Lewis acids (MX_n = TMSI, TMSBr, BBr_3, R_2Br, $AlBr_3$, AlI_3, or transition metal halides) liberate halide ion on complexation which completes the cleavage process. The intermediate metal alkoxides are covalently bound which reduces their tendency to participate in the reverse process—methylation—by the liberated methyl halide. In general:

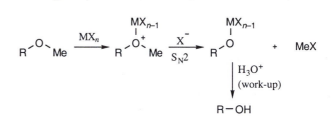

Deprotection of methyl ethers in the presence of other acid-labile PGs is difficult; certainly most acetals cleave under these conditions, but esters may be tolerated. The examples below also show that anchimeric assistance by a proximal hydroxy group greatly increases the rate of demethylation (see margin; complexation and activation are favoured by chelate formation).

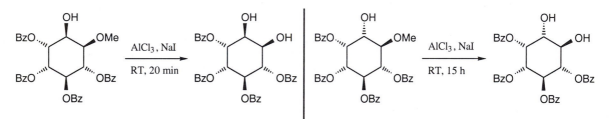

Both aryl and alkyl methyl ethers can be deprotected by similar hard* Lewis acid/soft nucleophile combinations but aryl methyl ethers are also labile to nucleophiles such as $LiPPh_2$, LiI, $NaCN$, and $RSNa$, where an effective Lewis acid is absent (Li^+ and Na^+ are *weakly* Lewis acidic). The ArO^- anion is a much more effective leaving group than RO^-, the negative charge being readily delocalised around the aromatic ring in the former case.

*Hard and soft acids and bases

	Acids	
	Soft	Hard
Size	Large	Small
Positive charge	Low	High

	Bases	
	Soft	Hard
Polarizability	High	Low
Electronegativity	Low	High
Easily oxidised?	Yes	No

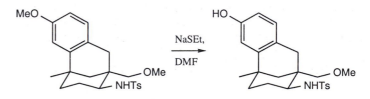

Conversely, conjugation of an oxygen lone pair in ArOMe renders the oxygen less Lewis basic so that *alkyl* methyl ethers, which lack this conjugation, can be selectively activated by Lewis acids and cleave more readily.

Codeine can be converted into morphine (structures, page 93) by O-demethylation with (BBr_3) or without (NaSPr) Lewis acid activation.

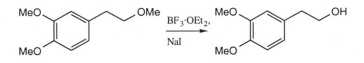

In summary:

	Lewis acid + nucleophile	Nucleophile in the absence of a good Lewis acid
Alkyl methyl ethers	faster (Lewis basicity dominates)	much slower
Aryl methyl ethers	slower	faster (leaving group ability dominates)

Substituted methyl ethers

Methyl ethers derive their stability from a reluctance to form CH_3^+ which is too unstable to have a significant lifetime in solution. Electron-donating substituents on the methyl group stabilise the positive charge and, when sufficiently stabilising, can initiate ionisation after protonation or complexation; those cases have been discussed already. However, there are intermediate cases that are *capable* of cleavage under protic hydrolysis conditions but are more easily cleaved by the Lewis acid/nucleophile combinations mentioned above. MOM and 2-(methoxyethoxy)methyl (MEM) ethers fall into this category, as do benzyl PGs for alcohols, carboxylic acids, and amines. The mechanistic principles are identical to those described for methyl ether cleavage except that, in these cases, there is more S_N1 character in the cleavage event as a consequence of electron release by either oxygen or a benzene ring. The cleavage step is concerted but asynchronous, i.e. there are no discrete intermediates but, in the transition state, bond-breaking is more advanced than bond-making and the leaving group will be detached before the newly-forming bond is completed:

See page 47 for a fuller discussion of 'synchronicity'.

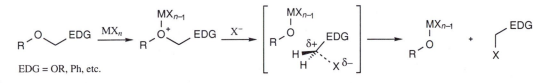

EDG = OR, Ph, etc.

The virtues of MOM protection have been covered on page 25; MEM ethers, introduced in a similar manner to MOM (e.g. NaH, KH, or *i*-Pr_2NEt + MEM-Cl), have a comparable stability towards protic acids, but are more reactive in the presence of Lewis acids. The presence of the extra oxygen atom favours Lewis acid complexation (*cf.* page 36) and cleavage with, for example, $ZnBr_2$, $TiCl_4$, R_2BBr, or TMSI is rapid; of course the acetal linkage also leads to MEM cleavage under protic hydrolysis conditions.

Coordination of Lewis acids (MX_n) with the MEM group.

Benzyl amines

Benzyl PGs for alcohols and carboxylic acids are usually selected because of their ability to be cleaved reductively, the Lewis acid/nucleophile cleavage pathway being available but less often used. However, benzyl *amines* are quite resistant to reductive cleavage but the nucleophilicity of the nitrogen atom enables a useful indirect deprotection method that is mechanistically related to the activation/substitution methods discussed so far. The von Braun reaction between 3°-amines and cyanogen bromide (BrCN)—developed for the degradation and characterisation of alkaloids—proceeds by cyanation of the amine followed by bromide attack at one of the alkyl groups liberating alkyl bromide. The reaction continues no further since the nitrogen atom is deactivated as a nucleophile, its lone pair of electrons being conjugated with the nitrile. The process is normally followed (dotted arrows) by hydrolysis of the nitrile group with aq. HBr to give a carbamic acid (which loses CO_2):

$$R_2\ddot{N}-C\equiv N$$

$$\updownarrow$$

$$R_2\overset{+}{N}=C=\overset{-}{N}$$

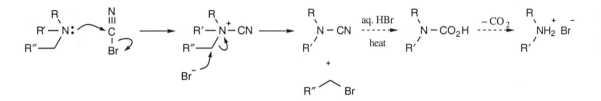

This chemistry has been 'modernised' in order to convert benzyl amines into carbamate derivatives, by an analogous mechanism, using the appropriate alkyl chloroformate, the specific choice of chloroformate being driven, as usual, by the subsequent constraints on PG tolerance and deprotection selectivity (see page 50).

i. Cl$_3$CCH$_2$OCOCl
ii. CH$_3$CH(Cl)OCOCl ('ACE-Cl')
iii. CH$_2$=CHOCOCl
iv. BnOCOCl
v. TMSCH$_2$CH$_2$OCOCl

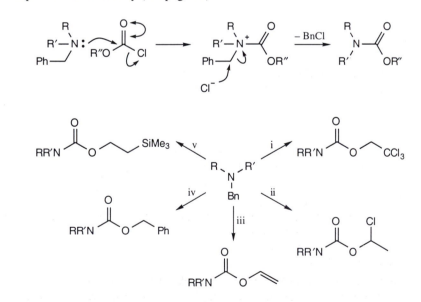

In example (1) the cyanobenzylidene functionality would not survive the reductive conditions usually used for benzyl group cleavage and example (2)

highlights the use of the trimethylsilylethyl carbamate PG which can be removed under mild conditions that leave acid-sensitive groups (e.g. EE) intact. See Chapter 3 for further discussion of these amine PGs.

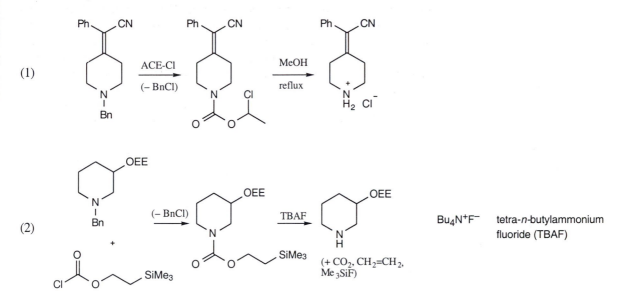

Other 3°-alkylamines may be cleaved reliably by this method, benzyl cleaving most easily, followed by allyl, followed by methyl.

As a footnote to this section, acid-labile PGs normally cleaved by protic hydrolysis are also likely to be reactive to Lewis acid/nucleophile combinations; the following list gives selected preferred reagents for deprotecting a variety of FGs:

t-Bu ethers	Ac_2O, $FeCl_3$
Tr ethers	$SnCl_2$ or $ZnBr_2$
Acetals	aq. $PdCl_2(CH_3CN)_2$ or R_2BBr
Boc-amines	TMSI or $AlCl_3$, anisole or BBr_3

PGs containing sulfur

Thiols are considerably less basic than alcohols (see margin) but, being less electronegative (O 3.5, S 2.5) and with valence electrons held further from the nucleus and therefore more polarisable, are much more nucleophilic ('soft', page 36). Therefore, cleavage by protic acid is usually ineffective (H^+ is a 'hard' electrophile) but soft Lewis acids, alkyl halides, halogens, and oxidising agents induce rapid activation.

Methylthiomethyl ethers

The methylthiomethyl (MTM) PG for alcohols (which, like most alcohol PGs, can also be used for phenols, carboxylic acids, thiols, etc.) is usually introduced in a similar manner to the MOM PG and imparts broadly similar levels of protection but the ease of oxidation of sulfur and its higher

$Bu_4N^+F^-$ tetra-n-butylammonium fluoride (TBAF)

ROH_2^+ $pK_a \approx -2$
RSH_2^+ $pK_a \approx -7$

pK_a is a thermodynamic measure, nucleophilicity is a kinetic property.

propensity to react with Lewis acids places limits on its use. The advantage of the MTM group is that it can be cleaved under almost neutral conditions and acid-sensitive FGs (such as *O,O*-acetals) remain unscathed. Deprotection can be achieved with aq. $HgCl_2$, aq. $AgNO_3$, or MeI/aq. $NaHCO_3$ mixtures, the mechanism of cleavage being comparable to that for *O,O*-acetal cleavage:

Formation and cleavage of MTM ethers.

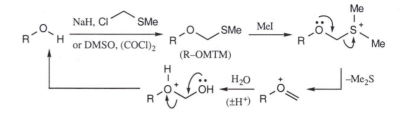

dimethylsulfoxide (DMSO)

2,6-lutidine

Two examples illustrate MTM deprotection in the presence of THP and acetonide PGs which are readily removed by protic hydrolysis. 2,6-Lutidine is added to prevent any possibility of competitive cleavage of the acid-labile PGs.

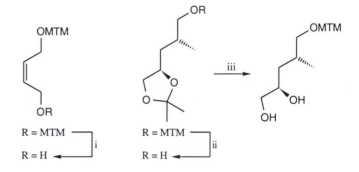

i. aq. $AgNO_3$, 2,6-lutidine
ii. aq. $AgNO_3$, $HgCl_2$, 2,6-lutidine
iii. aq. AcOH

Dithioacetals

The reactivity of oxathiolanes parallels that of *S,S*-acetals (see below).

MTM ethers are *O,S*-acetals, an FG that can also be used to protect carbonyl compounds by analogy to the formation of *O,O*-acetals. Of more importance are *S,S*-acetals (dithioacetals) which, by analogy to *O,O*-acetals, are usually prepared from the carbonyl compound with two equivalents of a thiol (or one of a dithiol) and an acid catalyst.

Protic and Lewis acids (particularly $BF_3 \cdot OEt_2$) promote dithioacetal formation, aldehydes being protected more rapidly than ketones and, at least in steroidal systems, carbonyl groups bearing α,β-unsaturation also react preferentially (contrast with acetalisation which takes place preferentially at the unconjugated carbonyl). Acyclic dithioacetals form and cleave more rapidly than their cyclic counterparts but selectivity for cleavage can only be achieved under carefully chosen conditions (e.g. aq. $GaCl_3$).

In our disussion of MTM ethers we noted that acetals containing sulfur are much less susceptible to protic hydrolysis and the equilibrium between ketone + thiol and dithioacetal + water lies almost completely in favour of the protected form (see OCP #33, page 7). In principle, no special steps need to be taken to remove water but in practice an excess of the Lewis acid 'catalyst' is usually present which also acts as a dehydrating agent.

Apart from their role in carbonyl protection, aldehyde dithioacetals are weakly acidic at the acetal carbon ($pK_a \approx 31$) and may be deprotonated by strong bases (e.g. BuLi), then alkylated. It is in this guise as an acyl anion equivalent that dithioacetals usually become incorporated into molecules but ultimately the deprotection considerations are identical. Below, in the synthesis of chalcogran, a Norwegian spruce beetle aggregation pheromone, various useful points are illustrated: (1) dithioacetals can be prepared directly from O,O-acetals, in this case a methyl furanoside; (2) protected aldehydes can serve as acyl anion equivalents by deprotonation and alkylation; (3) 1,3-dithianes can be converted into dimethyl acetals in the presence of (protic) acid-sensitive groups (the *soft* Lewis acid Hg(II) activates only the *soft* sulfur atoms of the dithiane, not the *hard* oxygen atoms in the THP acetal groups); (4) internal acetals (spiroacetals) are generated when suitable dihydroxyketone functionality is liberated, in this case by acidic hydrolysis in the final step.

$BF_3 \cdot OEt_2$ forms hydrates [$BF_3 \cdot OH_2$, $BF_3 \cdot (OH_2)_2$] in the presence of small amounts of water; when water is present in excess, disproportionation occurs:

$$4BF_3 + 6H_2O$$

$$\downarrow$$

$$3BF_4^- + B(OH)_3 + 3H_3O^+$$

2,4,6-collidine

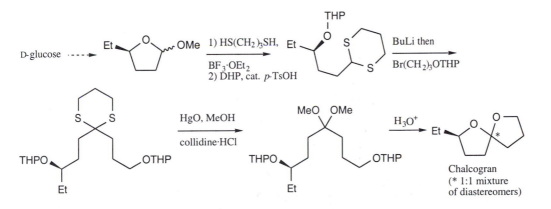

This example also shows that a heavy metal salt—usually Hg(II) or Ag(I)—is generally required* to induce cleavage; although these metal salts effectively complex the liberated thiol groups preventing re-addition, they are toxic and present disposal problems. Alternatives parallel those used for cleavage of MTM ethers, and alkylation (MeI or $Me_3O^+BF_4^-$) is often used. Oxidising agents and halogens are also effective when other FGs in the molecule can tolerate their presence. These reagents all work on the same mechanistic principle: conversion of one of the sulfur atoms into a better leaving group initiates sulfonium ion formation and hydrolysis.

RO–THP (page 26)

*Once more, H^+ is ineffective; refer to the discussion in the introduction to this section, page 39.

Metal ion method

A base may be added to suppress side-reactions caused by the liberated HX.

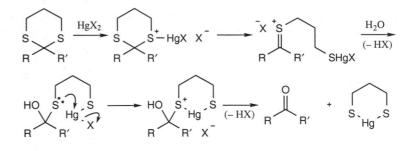

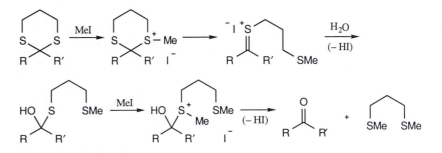

Alkylation method – parallels the MTM deprotection chemistry

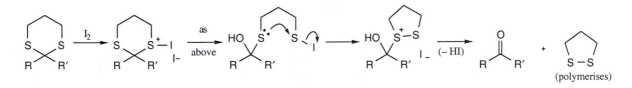

A similar mechanism can be written for deprotection with halogens or halogen transfer reagents (e.g. *N*-chlorosuccinimide),

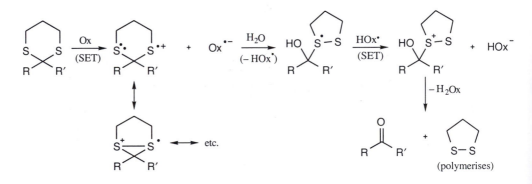

(polymerises)

* E.g. Ce(NH₄)₂(NO₃)₆   (CAN)
PhI(OCOCF₃)₂   (PIFA)
2,3-Dichloro-5,6-dicyano-1,4-benzoquinone   (DDQ)

but in this and many other deprotections that use oxidising agents* single electron transfer (SET) pathways are almost certainly followed, the basic idea still being the same. In the scheme below, 'Ox' is a generalised oxidising agent.

Whatever the detailed mechanism of these processes, the result is that *O,O*-acetals remain unaffected because they are not sufficiently nucleophilic to be activated by alkyl halides, nor are they prone to oxidation. A single example illustrates the selectivity in a complex molecule, the TBS, SEM, Bn, and acetonide PGs remaining intact.

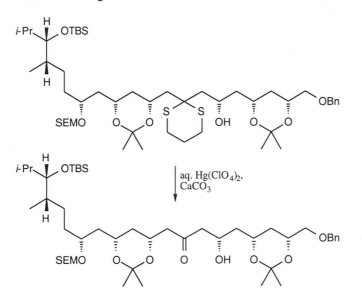

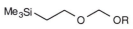 

2-(trimethylsilyl)ethoxymethyl ether (SEM-OR)

2.4 Deprotection following activation by alkylation

Pentenyl ethers, acetals, and related PGs

Acetal PGs cleave following oxonium ion formation initiated by protonation or attack by a Lewis acid. Oxonium ion formation may also be initiated by attack of one of the acetal oxygens onto an electrophile tethered* to the PG. This forms the basis of a small subset of PGs built around the pentenyloxy substructure to protect alcohols, carbonyl compounds, glycosidic positions, and amines. For example, pentenyloxymethyl (POM) ethers derived from alcohols can be deprotected selectively by treatment with halogens or their equivalents *N*-bromo- and *N*-iodosuccinimide (NBS, NIS), activation proceeding by kinetically favourable formation of a five-membered ring.

*Oxygen is less nucleophilic than sulfur but reliable activation by alkylation can be initiated by an intramolecular electrophile.

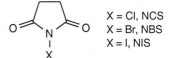

X = Cl, NCS
X = Br, NBS
X = I, NIS

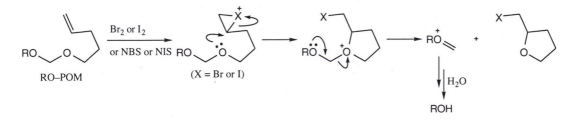

Pentenyl dioxolane acetal-protected carbonyl compounds are deprotected analogously

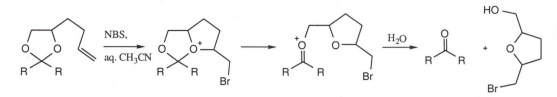

as are pentenyl glycoside derivatives of carbohydrates.

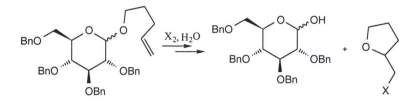

A variant of this idea is applied to amine protection. Thus 4-pentenamides can be deprotected selectively with I_2 in wet THF even in the presence of esters (which are more prone to acidic or alkaline hydrolysis).

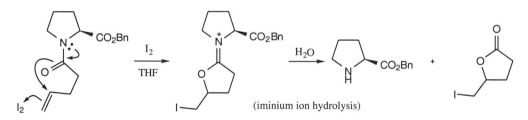

(iminium ion hydrolysis)

All of these derivatives provide the expected level of protection but offer an advantage over analogous PGs of possessing this extra Achilles' heel.

Hidden pentenyloxy functionality

The pentenyloxy grouping may not be obvious but may be buried within a molecule and lead to surprising levels of deprotection selectivity. In the example below only the 2-OH group is liberated by treatment with I_2 followed by Zn/AcOH, the others, including the (1°) 6-OH group, remain intact.

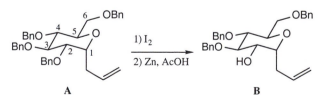

This result is closely related to deprotection of POM-protected alcohols. Thus, iodonium ion formation is followed by rapid trapping by the oxygen at the 2-position forming an oxonium ion. Loss of the benzyl group is assisted by nucleophilic attack by I⁻ ion and the resulting β-haloalkoxy arrangement is disposed for deprotection by β-elimination (Chapter 3). Taking advantage of FG proximity in this way adds yet another level of selectivity to the masking and unmasking of reactive FGs.

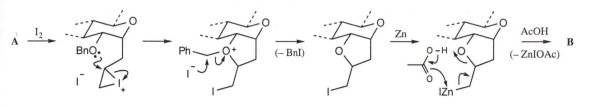

Further reading

Cited Oxford Chemistry Primers: #33: G. H. Whitham, *Organosulfur Chemistry;* #36: A. J. Kirby, *Stereoelectronic Effects;* #54: M. Grossel, *Alicyclic Chemistry;* #63: G. Procter, *Stereoselectivity in Organic Synthesis.*

3 Nucleophile/base-labile protecting groups

3.1 Introduction

The majority of the FGs described in this book are based on heteroatoms—O, N, and S—that, by virtue of higher electronegativity (O, N) or more diffuse orbitals (S), can support a negative charge more effectively than carbon. Consequently, some PGs can be cleaved by expulsion of the FG as an anion either by elimination (Section 3.4) or by nucleophilic substitution. Concerted (S_N2-like) substitution processes can be divided, somewhat arbitrarily, into *pure* substitutions, where the leaving group is ejected without strong prior activation,* and *assisted* substitutions where nucleophilic displacement of the FG requires significant activation by protonation or Lewis acid complexation (Chapter 2).

Whether or not activation is necessary depends upon (a) the leaving group ability of the protected FG—thus carboxylic acids and the carboxyl group in carbonates and carbamates may be liberated without significant activation, (b) the nucleophilicity of the attacking species—soft (page 36), poorly solvated nucleophiles are the most reactive, (c) the solvent—polar, aprotic solvents (pyridine, DMF, DMSO) leave anionic nucleophiles uncluttered by solvation spheres, and (d) the steric availability at the electrophilic site—1°- and unhindered 2°-alkyl substituents support S_N2 chemistry.

'Electronegativity is the power of an atom in a molecule to attract electrons to itself,' Pauling, 1932.

| C | 2.5 | N | 3.0 | O | 3.5 |
| Si | 1.8 | | | S | 2.5 |

Pauling electronegativity values.

*Anions are stabilised in solution by solvation (especially in protic solvents by H-bonding) and by complexation to a metal counterion.

dimethyl formamide (DMF)

dimethyl-sulfoxide (DMSO)

3.2 Protecting group cleavage by S_N2 chemistry

Carboxylic esters

Methyl esters (preparation: page 56) contain an inert stand-in in place of the acidic carboxyl proton but offer little protection of the carboxyl carbon towards nucleophilic addition ($B_{Ac}2$ methods: Section 3.3) and are, as a consequence, easy to deprotect when necessary. Because carboxylate anions are reasonable leaving groups, and methyl is the smallest alkyl group, deprotection can proceed by a $B_{Al}2$ mechanism:

$B_{Al}2$: ester hydrolysis under <u>b</u>asic conditions with cleavage of the methyl (<u>al</u>kyl) oxygen bond in a <u>b</u>imolecular rate-determining step. Subset of S_N2. (Metal counterion omitted for clarity.)

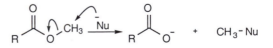

For optimum yields the nucleophile is usually I^- (as LiI), or CN^-, or a thiolate (RS^-), and the solvent a polar aprotic one (listed above). This method for regenerating a carboxylic acid from an ester is essentially neutral—no strong acid or base is needed—and very little water is necessary (just sufficient to protonate the carboxylate) which leads to excellent selectivity in favour of methyl esters. The examples below illustrate the effect of steric hindrance in reducing the rate of $B_{Al}2$ cleavage as neither the ethyl nor the phenyl esters are affected.

As well as the steric availability of the alkyl group in determining the rate of $B_{Al}2$ cleavage, electronic effects are also important.

Phenacyl esters, formed from carboxylic acids with phenacyl bromide and a base (Et_3N or KF), are stable to acidic hydrolysis but may be removed at room temperature by treatment with NaSPh or PhSeH in DMF via a $B_{Al}2$ pathway. Benzyl esters, too, *can* be removed by unassisted nucleophilic substitution at the benzyl CH_2 but the conditions required are more forcing (NaHTe in hot DMF) and other assets of the benzyl group are usually exploited in deprotections (Chapter 5).

Although reasonably bulky, both phenacyl and benzyl esters benefit from a π-system adjacent to the site of substitution, a feature which markedly increases their reactivity towards substitution. This effect originates in the fact that concerted S_N2 substitutions are rarely *synchronous,* i.e. the extent of bond formation (to the nucleophile) does not necessarily proceed at exactly the same rate as bond cleavage (to the leaving group). If the incoming and outgoing groups differ in their capacity to support a negative charge then the lowest energy transition state will reflect an optimised, unsymmetrical, distribution of electrons; i.e. it will be polarised to some degree.

Either side of an idealised concerted, synchronous S_N2 reaction lie points on the mechanistic spectrum between S_N1 (leaving group loss precedes addition of the nucleophile) and substitution by an addition–elimination pathway (*cf.* cleavage of Si–C bonds via a siliconate intermediate); both of these extremes are non-concerted. The mid-point for each mechanistic type (intermediate or transition state) is summarised below:

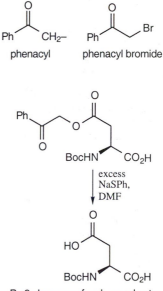

phenacyl   phenacyl bromide

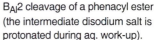

$B_{Al}2$ cleavage of a phenacyl ester (the intermediate disodium salt is protonated during aq. work-up).

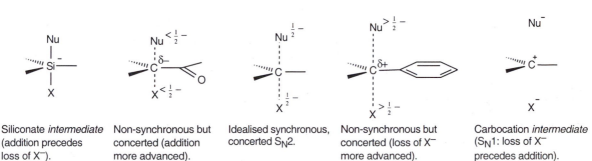

| Siliconate *intermediate* (addition precedes loss of X⁻). | Non-synchronous but concerted (addition more advanced). | Idealised synchronous, concerted S_N2. | Non-synchronous but concerted (loss of X⁻ more advanced). | Carbocation *intermediate* (S_N1: loss of X⁻ precedes addition). |

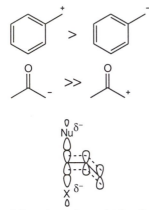

Adjacent π-systems stabilise the S_N2 transition state by overlap with the Nu--C--X orbitals.

*Rate of $S_N2 \propto$ [Nu][substrate]

From these diagrams it can be seen that an electron-rich aromatic π-system stabilises partial positive charge at the site of attack (favouring an asynchronous pathway tending towards S_N1) and an electron-poor carbonyl π-system will stabilise a build-up of negative charge in the transition state (favouring an asynchronous pathway tending towards addition–elimination).

Mesomeric stabilisation by the phenacyl carbonyl group of the (electron-rich) S_N2 transition state is supported by an electron-withdrawing inductive effect. This –*I* effect also raises the δ^+ charge on the CH_2 carbon which increases the electrostatic attraction of that position towards nucleophiles. Benzyl esters lack these inductive contributions and favour cleavage conditions that promote S_N1 (page 21).

Intramolecular substitution

Methyl esters are unhindered and phenacyl esters are electronically activated towards S_N2 cleavage but the reaction is less effective with ethyl esters and is not synthetically useful with most higher esters. However, if the deprotective substitution step is *intramolecular* the effective concentration of the nucleophile is increased* and the rate is raised sufficiently so that bulkier esters, offering increased steric protection of the carboxyl carbon, may be removed by treatment with a good nucleophile. For example, 4-chlorobutyl and 5-chloropentyl esters react with Na_2S to give a potent sodium alkylthiolate nucleophile perfectly disposed for intramolecular S_N2 displacement of the carboxylate leaving group, e.g.

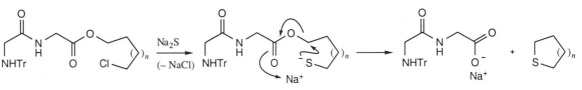

AcO–H $pK_a = 4.76$
PhO–H $pK_a = 10.0$
CyO–H $pK_a \approx 16$

The PG chemistry of phenols is intermediate to that of carboxylic acids and alcohols; whilst the phenoxide ion is stabilised by conjugation with the aromatic ring this is not sufficient to confer much leaving group ability and cleavage of most phenol PGs requires Lewis acid activation (but see page 36).

Sulfonium salts

The strategies for reducing the nucleophilicity of dialkyl sulfides (R_2S) are analogous to those for 3°-amines (R_3N): salt formation (R_3S^+ *cf.* R_4N^+, page 5) and oxidation (R_2S^+–O^- *cf.* R_3N^+–O^-, page 93).

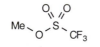

methyl triflate (MeOTf)

Methylation of dialkyl sulfides (R_2S) with, for example, MeOTf deactivates the nucleophilicity of the sulfur atom and, because the methyl group provides little steric interference towards nucleophilic attack, discrimination is possible between the three groups attached to sulfur during eventual deprotection (with $HOCH_2CH_2SH$ or $LiAlH_4$).

However, if, instead of methyl, an alkyl group with a latent internal nucleophile is chosen, the deprotection step is unambiguous. For example, alkylation of dialkyl sulfides with 4-bromobutylphthalimide achieves protection of the sulfur atom against further alkylation or oxidation; when no longer required, the PG can be removed by the addition of a large excess of aq. $MeNH_2$ (*cf.* page 59). This liberates a 1°-amine which cyclises to generate pyrrolidine and the free dialkyl sulfide in a process analogous to chloroalkyl ester deprotection. Thus PGs can be 'nested', deprotection of one FG leading to a further reaction that exposes a second FG. The scheme below illustrates protection and deprotection of a simple dialkyl sulfide with the 4-(phthalimido)butyl group.

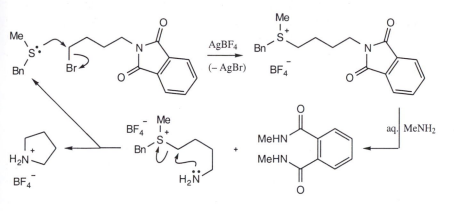

3.3 Protecting group cleavage by nucleophilic addition

The ability of the carbonyl carbon to accept a nucleophile prior to fragmentation of the so-formed tetrahedral intermediate underlies the cleavage of a diverse range of PGs. Addition of a nucleophile to a generalised carbonyl compound is reversible if Nu⁻ can compete with B⁻ as a leaving group and the sequence may continue by further additions of Nu⁻ ($\rightarrow$ Nu$_2$C=O, etc.).

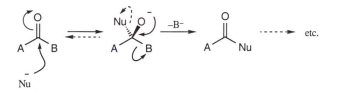

The case A = alkyl, B = OR corresponds to B$_{Ac}$2 cleavage of an ester.

B$_{Ac}$2: basic, acyl-oxygen fission, bimolecular

The overall result—acyl transfer from B to Nu—corresponds to protection of Nu and deprotection of B; therefore, the mechanisms underlying acyl PG introduction and cleavage are identical.

Acylation of nucleophilic FGs (alcohols, amines, thiols) reduces the electron density on the nucleophilic atom (O, N, S) by conjugation (Section 1.3). Conversely, conversion of a carboxylic acid into an ester or amide removes side-reactions attributed to the acidic proton and provides shielding against the attack of nucleophiles at the carboxyl carbon. Esters can provide a steric barrier against nucleophilic attack; the protection offered by amides is largely electronic in origin.

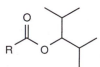

The carboxyl carbon in diisopropyl-methyl (diisopropyl carbinyl) esters is shielded from external nucleophiles.

Nucleophilic addition to amides results in loss of conjugation which carries a significant energetic penalty.

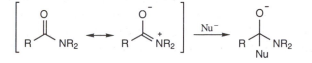

Acylation of hydroxyl groups

Acylation of alcohols (and amines and thiols) is usually achieved using the appropriate acyl chloride (or alkyl chloroformate) in the presence of a base to scavenge liberated HCl, or with the carboxylic acid itself in the presence of an activator (e.g. DCC) and a nucleophilic catalyst such as DMAP. Acylation with DCC/DMAP consists of two phases: (i) activation of the carboxylic acid, then (ii) acylation. The first phase is invariant for esters and amides:

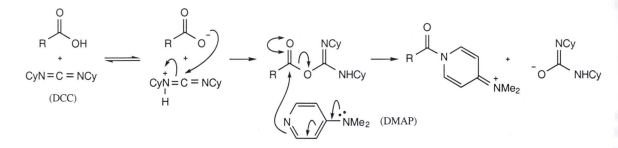

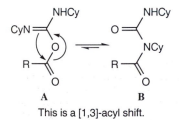

A   **B**

This is a [1,3]-acyl shift.

DMAP is regenerated after the acylation step and only a catalytic quantity is required. The hydroxyl proton is removed to form dicyclohexyl urea (DCU).

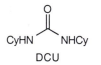

DCU

*Of cyclohexyl alcohols those occupying equatorial sites are more reactive than those in axial sites.

R' = H   formate
R' = CH₃   acetate (ROAc)
R' = Ph   benzoate (ROBz)
R' = *t*-Bu   pivaloate (ROPv)

DMAP is a nucleophilic catalyst in this reaction; being more nucleophilic than most alcohols and amines it reacts rapidly with the carboxylic acid + DCC adduct to give a reactive extended acyl iminium electrophile. This prevents the isourea intermediate (**A**, in margin) from rearranging to the more stable *N*-acyl urea (**B**) and produces a more reactive acylating agent which is then trapped in the second part of the reaction to give the ester or amide product. The process is illustrated for esterification:

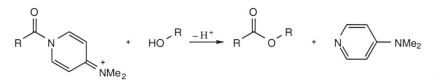

Steric hindrance is a significant determinant of acylation selectivity and, in general: (a) 1°-alcohols can be acylated more rapidly than 2°-alcohols, (b) of 2°-alcohols those in less-hindered positions are more reactive,* (c) 3°-alcohols acylate only slowly and very reactive acylating agents are usually needed, and (d) bulky acylating agents exaggerate these trends. These general trends apply also to the acylation of other FGs.

Ester derivatives of alcohols

With the proviso of point (d) above, most esters may be prepared from a particular alcohol with broadly equal selectivity; therefore, the major factors that guide the choice are (i) the tolerance of the ester to the intended chemical steps and (ii) its ability to be cleaved in the presence of other functionality.

Formate esters are normally too unstable with respect to hydrolysis to be of significant use as PGs (*ca.* 100× more reactive than acetates). Acetates tolerate a wide range of chemical procedures but strong nucleophiles (certain organometallic reagents, metal hydrides, amines) or combinations of an acidic reagent with either water or an alcohol must be avoided. They find most use in carbohydrate and peptide synthesis where more than one hydroxyl group often needs to be protected and the act of protection has to be highly efficient at each hydroxyl group. Aside from certain limitations in chemical tolerance, acetates are prone to migration from one position to another; complex mixtures may be obtained unless there is a distinct difference in stability between the possible products. Migration of acetate from the 4- to 6-OH in glucose derivatives is a reliable reaction that exposes just the 4-OH group.

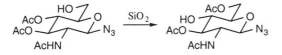

Benzoate esters are used almost as frequently as acetates, partly because they are much less prone to migration yet offer similar levels of protection. Both acetate and benzoate esters may be cleaved with nucleophilic reagents such as NH_3, alkylamines, or K_2CO_3 in MeOH, and alkali metal hydroxides in mixed aqueous–organic solvents are also effective ($B_{Ac}2$ mechanism).

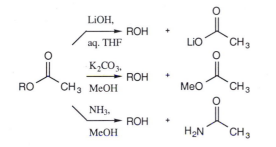

Ester cleavage by hydrolysis or by addition of an amine results in a stable carboxylate or amide respectively; base-catalysed methanolysis relies on a large excess of MeOH (the solvent) to favour the cleavage.

However, there are subtle differences in reactivity that enable acetate and benzoate protected alcohols to be differentiated. The carbonyl group in benzoate esters is conjugated with the phenyl ring so that nucleophilic addition is more difficult than to acetates and selective cleavage of the latter is possible. Furthermore, the CH_3 protons in acetate esters are mildly acidic ($pK_a \approx 24$) which leads to a sensitivity to base; cleavage can be effected with DBU under aprotic conditions that leave non-enolisable esters intact:

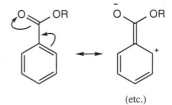

(etc.)

via:

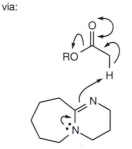

diazabicycloundecene (DBU)

Enzymatic cleavage can lead to levels of selectivity unsurpassed by conventional laboratory reagents. Although this is not usually the first choice method, cleavage of just one of two acetates in a *meso*-diacetate results in a single enantiomer, as illustrated by the preparation of starting materials for prostaglandin synthesis, the choice of esterase providing either enantiomer of the monoacetate.

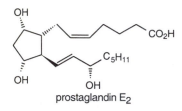

prostaglandin E₂

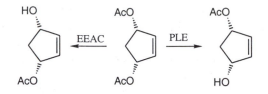

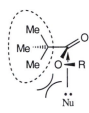

PLE   pig (porcine) liver esterase

EEAC   electric eel
       acetylcholinesterase

In, general $k_2 > k_1$ and 2 equivalents
of MeLi are necessary even if the
reaction is kept cold (−78°C).

Higher in stability is the pivaloate derivative, the stability originating in the steric impediment to approach of a nucleophile to the carbonyl carbon. The hydrolysis conditions are sufficiently harsh that, for example, silyl ethers (Chapter 4) elsewhere in the molecule may be cleaved (in these cases non-hydrolytic conditions are favoured, see below). Because cleavage and acylation are mechanistically related, relatively unhindered 1°-alcohols are pivaloated more readily.

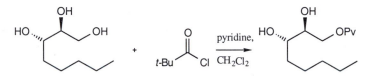

In substrates that are prone to base-induced epimerisation or contain alkoxide-labile functionality, MeLi or hydride reducing agents (DIBAL or LiAlH₄) are preferred. The cleavage mechanism parallels B$_{Ac}$2 hydrolysis; for each deprotection two nucleophilic additions are necessary:

Beyond these three ester PGs the others are mere variations, the effects of additional substitution on the ease of cleavage being predictable. Electron-withdrawing groups enhance the reactivity of the ester carbonyl towards cleavage (and vice versa), and increasing steric bulk in the proximity of the carboxyl carbon reduces it (*cf.* Ac and Pv esters).

These points are illustrated in the stereocontrolled preparation of both the diastereomers (**A** and **B**, below) of a monosilylated alkenyl diol. The *para*-nitrobenzoyl (PNBz) and mesitoyl (Mes) esters are used respectively as enhanced and deactivated versions of Bz.

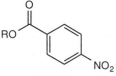

RO–PNBz

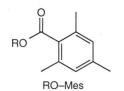

RO–Mes

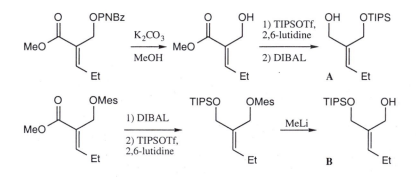

Key points to note are as follows: For **A**: (1) the *para*-NO$_2$ group is powerfully electron withdrawing (inductive and mesomeric) and raises the reactivity of the PNBz carboxyl carbon above that of the methyl ester; (2) the use of K$_2$CO$_3$ in MeOH ensures that any attack at the methyl ester is a degenerate process (substitution of MeO for MeO); (3) the bulky triisopropylsilyl (TIPS) ether (page 78) ensures that no migration of the silyl group occurs during the reduction step.* For **B**: (1) the *ortho*-methyl substituents in the Mes ester provide a very effective block towards addition to that carbonyl and DIBAL reduction ensues solely at the methyl ester; (2) because the Mes ester is very hindered MeLi is required to deprotect it.

*DIBAL (*i*-Bu$_2$AlH) is the preferred reagent for reducing α,β-unsaturated esters to allylic alcohols.

Attenuation of reactivity is also shown in the rates of hydrolysis of the chloroacetates which increase in response to progressive destabilisation of the carbonyl group by the inductive effect of the chlorine atoms.

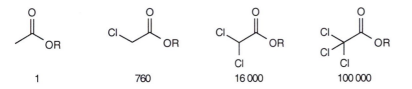

Relative rates of alkaline hydrolysis.

| 1 | 760 | 16 000 | 100 000 |

Of these, the most useful is the monochloroacetate as this can be cleaved under non-aqueous conditions because the carbonyl group enhances the rate of substitution at the α-centre (page 47) by soft (page 36) sulfur and nitrogen nucleophiles that favour substitution at the softer α-centre.

With suitable ambient* nucleophiles S$_N$2 substitution brings a second nucleophilic component into close proximity to the ester group, raising the effective concentration of the reagent at the carbonyl carbon and enabling good selectivity between esters.

*Ambident nucleophiles are nucleophilic at more than one atom; examples include CN$^-$ and enolate ions.

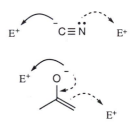

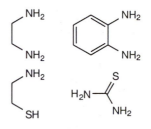

In this context ambident nucleophiles are those that are capable of two sequential substitutions from two different positions and include diamines and thioamines, e.g.

Displacement of the α-chlorine by the sulfur atom precedes lactam formation and expulsion of the alcohol without affecting the acetate.

Levulinate esters (ROLev), incorporate a keto group within the carboxyl component to allow a comparitively slow addition (to ester) to be replaced by a faster addition (to ketone) followed by a fast cyclisation. A similar idea is used in the *reductive* deprotection of Lev esters (page 92).

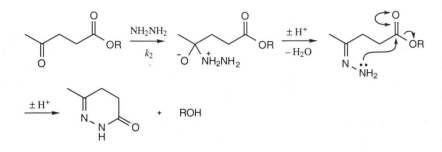

Formation of the tetrahedral intermediate is rate determining in addition of hydrazine to esters.

In general $k_2 > k_1$ (no disruption of conjugation in the case of addition to ketones).

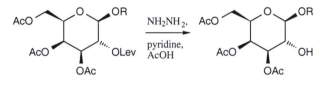

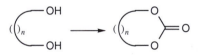

Lev-protection of the 2-OH group in a galactose derivative allowed selective cleavage, in the presence of three acetate esters, later in the synthesis by addition of NH_2NH_2:

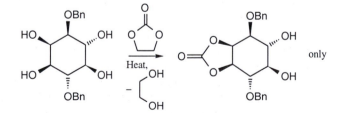

Protection of diols as cyclic carbonates

Although spacially separated hydroxyl groups within the same molecule can be capped with the requisite number of ester groups, it is often desirable to identify any 1,2- or 1,3-diol pairings that may be tied together selectively as a cyclic carbonate leaving other hydroxyl groups unaffected.

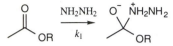

Ring-size nomenclature

Small   3, 4
Normal   5, 6, 7
Medium   8, 9, 10, 11
Large   12 and above

In acyclic systems, in principle, any 1,2- or 1,3- diol pairing may be protected in this way. However, in cyclic systems (especially normal rings), a further level of selectivity can be achieved since it is difficult to form a *trans*-fused bicyclic structure when one of the rings is five-membered and only *cis*-1,2-diol pairings react rapidly in this reaction. Also, carbonate formation is selective for *cis*-1,2-diols in preference to *cis*-1,3-diols as the latter lead to more strained bridged bicyclic structures, e.g.

The *cis*-1,2-diol is protected in preference to both the *cis*-1,3-pair and the *trans*-1,2-pair.

Originally, carbonate formation was achieved with phosgene (+ pyridine) but this is a highly toxic gas and diphosgene (liquid), triphosgene (solid), or phosgene equivalents (carbonyl diimidazole, ethylene carbonate) are now used.

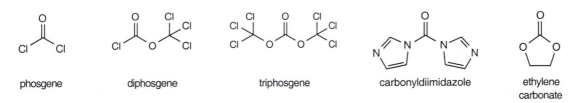

phosgene   diphosgene   triphosgene   carbonyldiimidazole   ethylene carbonate

Triphosgene is the reagent of choice for protecting *cis*-1,3-diols in molecules where no *cis*-1,2-relationship exists, e.g.

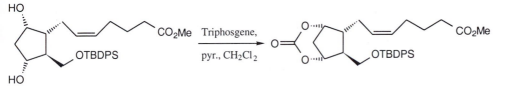

Carbonate C=O groups are conjugated with two oxygen atoms and are less reactive than ester C=O groups towards nucleophiles. Nevertheless, cyclic carbonates are readily removed with metal hydroxides in aqueous solution, or amine bases (NH_3, pyridine) in aqueous or alcoholic solution, but they are very stable towards acidic hydrolysis conditions which is useful in carbohydrate chemistry to protect diols during acetal hydrolysis or glycosylation processes. In the example below the carbonate protects the *cis*-diol during PDC oxidation of the unprotected OH group, survives the Baeyer–Villiger oxidation, but is removed reductively by DIBAL:

t-BuPh₂Si-   *tert*-butyldiphenylsilyl (TBDPS)

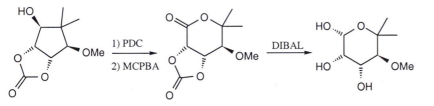

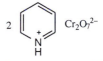

pyridinium dichromate (PDC)

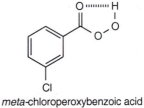

meta-chloroperoxybenzoic acid (MCPBA)

Esters for carboxyl protection

If mere masking of the acidity of a carboxylic acid is required and there are no particular limitations on PG removal, a simple alkyl ester will suffice (methyl, ethyl), but if it is necessary to protect against nucleophilic attack or the substrate is sufficiently complex that ester cleavage has to be unambiguous then special esters will be required. Of the latter, *tert*-butyl and benzyl esters are the most commonly used; the *tert*-butyl group offers moderate protection against nucleophilic attack and is readily removed under acidic conditions (page 19), the benzyl group is usually removed by selective hydrogenolysis (page 89) or dissolving metal reduction (page 67). Most of the other variations provide predictable deprotection selectivity as discussed in the context of alcohol protection.

The major methods for the preparation of alkyl esters are summarised overleaf. Choosing between these is easy: pick the simplest available method that is compatible with the other functionality in the molecule.

*Oxalyl chloride (ClCOCOCl) + cat. DMF (HCONMe₂) is a convenient alternative to thionyl chloride (SOCl₂).

EtO₂C-N=N-CO₂Et
diethylazodicarboxylate (DEAD)

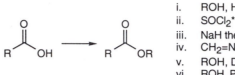

i.   ROH, H⁺ (– H₂O)
ii. SOCl₂* then ROH, Et₃N
iii. NaH then R-I
iv. CH₂=N₂ or TMSCH=N₂ (methyl esters)
v.   ROH, DCC, DMAP
vi. ROH, PPh₃, DEAD (Mitsunobu reaction)

As a corollary of offering fairly low levels of protection the lower alkyl esters—particularly methyl esters—may be removed under relatively mild conditions, e.g. alkali metal hydroxides or carbonates in aqueous or alcoholic solution. These unhindered alkyl esters may also be cleaved by nucleophilic substitution (B_{Al}2, page 46), a process that is ineffective for higher alkyl esters and, as with acetate derivatives of alcohols, enzymatic hydrolysis may also be used if other methods fail.

For example, in attempts to produce an acrylate precursor to a polymeric adhesive, chemical methods failed to hydrolyse the methyl ester efficiently because the acrylate was attacked competitively. Fortunately, an enzyme, the lipase from *Candida cylindracea,* achieved the hydrolysis in 89% yield.

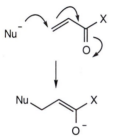

Vinyl carbonyl compounds are prone to 1,4-addition at the unhindered alkene terminus which, under basic conditions, can lead to polymerisation.

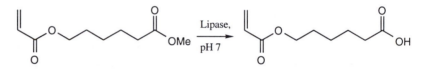

Intramolecular esterification (lactonisation) of a carboxylic acid by electrophile-induced cyclisation onto a suitably disposed alkene offers an alternative PG scheme because there are reliable procedures for regenerating both the the acid and the alkene when protection is no longer required. (Viewed another way, the -CO₂H group can be used to protect an alkene, *cf.* page 9). Iodolactonisation can differentiate the two carboxyl groups since only that which can approach the alkene (→ iodonium ion) can cyclise, leaving the other free to be esterified. At the end of the sequence both alkene and acid can be regenerated by Zn-mediated β-elimination (page 65) which leaves the methyl ester intact; the overall result is selective esterification of just the *exo*-carboxylic acid.

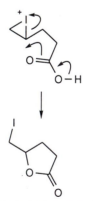

Iodolactonisation proceeds via trapping of an iodonium ion by a proximal -CO₂H.

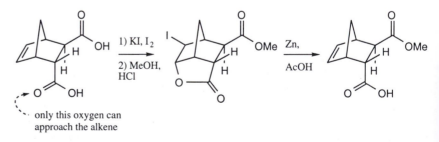

only this oxygen can approach the alkene

Release of the free acid from a lactone by nucleophilic addition leaves an OH group where the alkene was, which, in the example, cyclises to form the *endo*-epoxide. Direct epoxidation of the alkene leads to the *exo*-epoxide.

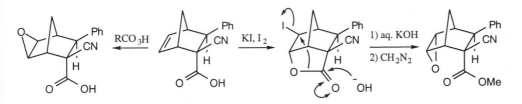

Amides for carboxyl protection

By analogy to mutual protection of alcohols and carboxylic acids as esters, amines and carboxylic acids may be mutually protected as amides. A vast array of methods is available for producing an amide from its constituent amine and acid because this is fundamental to the construction of peptides. Methods parallel those used to form esters with emphasis on either amine plus acid chloride or direct coupling methods between amine and the acid.

$$\underset{R}{\overset{O}{\|}}{\text{OH}} \quad + \quad R_2NH \quad \xrightarrow[\text{reagent(s)}]{\text{coupling}} \quad \underset{R}{\overset{O}{\|}}{NR_2}$$

Amide formation offers a route to protect both carboxylic acids and amines. Many coupling reagents are described in OCP #7.

Amides offer substantial protection of carboxyl groups against nucleophilic addition (pages 5 and 49) but often require such harsh cleavage conditions that their use in PG chemistry has been limited to simple substrates where strongly basic or acidic conditions are tolerated. However, 3°-amides may be cleaved selectively in the presence of 1°- or 2°-amides using *t*-BuOK in aq. Et$_2$O at room temperature (only very hindered amides require heating). Under these conditions a thermodynamically unfavourable addition of HO$^-$ to the amide is probably followed by deprotonation to the dianion which then collapses irreversibly to liberate the carboxylate and amide (which then protonates). 1°- and 2°-amides (that bear an N-proton) are deprotonated under these conditions to form a carboxamide salt which is unreactive towards addition of OH$^-$.

$pK_a \approx 17\text{–}18$

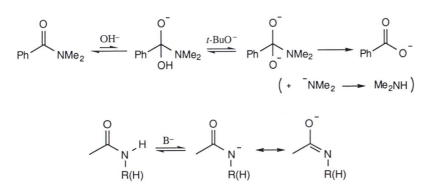

$$\left(+ \ ^-NMe_2 \longrightarrow Me_2NH \right)$$

B$^-$ = OH$^-$ or *t*-BuO$^-$

Direct hydrolysis of 1°- and 2°-amides is problematic but methods have been devised that activate the amide prior to cleavage. For example, preparation of the *N*-Boc derivative reduces the extent of conjugation of the nitrogen atom into the amide carbonyl such that cleavage can be achieved using conditions comparable to those used for ester hydrolysis, e.g.

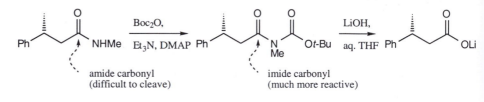

amide carbonyl
(difficult to cleave)

imide carbonyl
(much more reactive)

Amides for amine protection

Although amides are of limited use for carboxyl protection, they have a central place in the PG chemistry of amines alongside carbamate derivatives and imides (see below). In general, the reactivity order towards nucleophilic cleavage of the most useful amides is trifluoroacetamide* (R_2NCOCF_3) > formamide (R_2NCHO) > acetamide (R_2NAc) > benzamide (R_2NBz).

Trifluoroacetamides can be selectively introduced onto 1°-amines

*The three fluorine atoms exert a powerful destabilising *–I* effect on the amide C=O carbon which raises its reactivity towards nucleophilic addition.

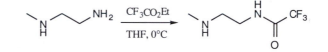

and may be deprotected under conditions where even esters remain unaffected (e.g. $NaBH_4$, EtOH) as well as those in which esters are cleaved competitively (K_2CO_3 or NH_3 in MeOH).

Formamide deprotection requires more stringent conditions (aq. HCl or aq. NaOH, reflux) and acetamides and benzamides need activation of the kind discussed above for carboxyl protection in order to avoid harsh conditions. Fortunately, enzymatic methods can be very useful here and acylases may be used to hydrolyse acetamides at essentially neutral pH; in example (1) the α-nitrogen is deprotected in the presence of the acetanilide* and example (2) shows removal of an acetamide PG in the presence of the much more reactive phthalimide (page 59). Both examples illustrate the enantiospecificity shown by enzymes, only one enantiomer being processed in each case as a consequence of the inherent chiral environment at the enzyme active site.

*Aryl acetamides are expected to be more reactive to chemical hydrolysis than alkyl acetamides. Why?

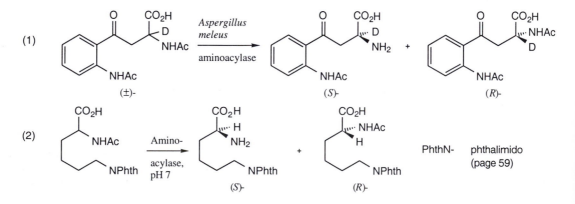

Phenyl acetamides may also be cleaved enzymatically, acid-sensitive acetals remaining unscathed:

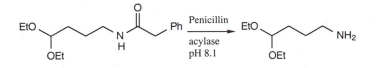

Imides for amine protection

Aminoacylases and special amide activation procedures have extended the use of amides as PGs, but the convenience and tolerance to oxidative conditions of *imides* have been long appreciated. The phthalimide (Phth) group ties together all of the bonding potential of a 1°-amine in a single group; the nitrogen lone pair is delocalised into two carbonyl groups and its stabilising effect on either one of them is roughly halved compared with that in amides.

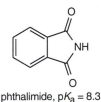

Imides are *N*-acyl amides.

phthalimide, pK_a = 8.3

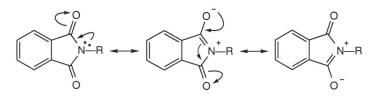

As a consequence phthalimides undergo nucleophilic attack relatively easily and, whereas they are stable to oxidising agents and most acidic reagents (but not forcing conditions, e.g. aq. H_2SO_4 at reflux), they are susceptible to attack by amines, organolithium, Grignard, and hydride transfer reagents. Hydrazine (NH_2NH_2) is ideal for cleaving the phthalimide because, after the first addition, the remaining carbonyl group is no longer part of an imide and is quite resistant to further *inter*molecular addition but the proximity of the second nitrogen sets up a fast cyclisation to complete the deprotection.

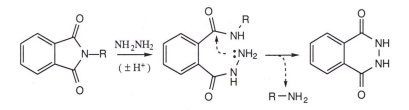

The phthalimide group can be introduced by nucleophilic substitution of an alkyl halide by the (potassium) salt of phthalimide.

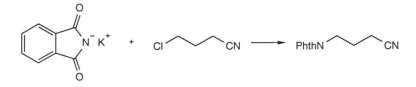

Combined with subsequent addition of NH_2NH_2 this constitutes the Gabriel synthesis of 1°-amines.

Alternatively, the Mitsunobu reaction can be used to convert 1°- and 2°-alcohols directly to -NPhth derivatives with inversion of configuration.

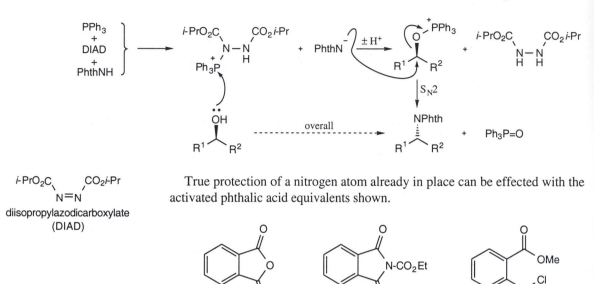

diisopropylazodicarboxylate
(DIAD)

True protection of a nitrogen atom already in place can be effected with the activated phthalic acid equivalents shown.

phthalic anhydride   *N*-ethoxycarbonyl phthalimide   methyl phthaloyl chloride

Protection of the NH$_2$ group in the indole derivative shown prevents it competing for the benzyl bromide with the inherently less nucleophilic indole nitrogen (the lone pair forms part of the aromatic system).

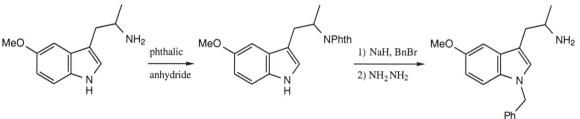

Carbamates for amine protection

Carbamates (urethanes) are *N*-(alkoxycarbonyl)amines. Urethane is NH$_2$CO$_2$Et.

An advantage of carbamate PGs is that the cleavage conditions can be varied widely depending on the choice of alkyl component. Thus benzyl carbamates (-NCbz) may be deprotected hydrogenolytically (page 89), allyl carbamates (-NAlloc) are susceptible to cleavage with Pd(0) catalysis (page 72), trichloroethyl carbamates (-NTroc) are deprotected with Zn/HCl (page 65), *t*-butyl carbamates (-NBoc) release the amine under acidic conditions (page 20), fluorenylmethyl carbamates (-NFmoc) are usually cleaved with amine bases (page 63), and trimethylsilylethyl carbamates (-NTeoc) undergo F$^-$ ion-induced fragmentation (page 68).

Of the carbamates that incorporate no such PG devices in the alkyl component only two find common application: methyl and ethyl. Since these lack a chemical Achilles' heel, forcing conditions are usually needed to cleave them.* Thus alkaline hydrolysis [KOH or Ba(OH)$_2$] usually requires heating to around 100°C (in aq. MeOH or ethylene glycol) but addition of NH$_2$NH$_2$ occurs at lower temperatures as does hydride reduction with Red-Al, although these conditions can interfere with other FGs such as esters, e.g.

*Nucleophilic attack at the carbamate carbon must disrupt the resonance stabilisation conferred by both the O and N atoms (see page 5).

NaAlH$_2$(OCH$_2$CH$_2$OMe)$_2$ (Red-Al)

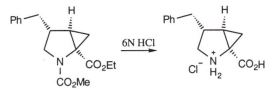

These limitations aside, methyl and ethyl carbamates are cheaply and easily prepared from the amine and the appropriate alkyl chloroformate in the presence of a base, offer good protection against electrophiles including oxidising agents, and reduce the ease of racemisation of α-amino acid derivatives.

N-Sulfonamides

Aryl sulfonyl derivatives of amines are susceptible to cleavage with Na in liquid NH$_3$ (page 93) but are otherwise fairly unreactive (which is why they are used !), and are often crystalline. In fact, sulfonamides of simple *alkyl* amines are rarely cleaved by nucleophilic addition chemistry because the conditions required are so harsh; LiAlH$_4$—a very powerful reducing agent— normally leaves aryl sulfonamides unaffected at temperatures much below *ca.* 100°C. Trifluoromethanesulfonamides, on the other hand, are sufficiently reactive to be cleaved by Red-Al at room temperature (the method of choice) or with LiAlH$_4$ at around 35°C as exemplified by the later stages in the synthesis of (+)-papuamine, the enantiomer of an antifungal natural product.

trifluoromethanesulfonamide
(TfNH$_2$)

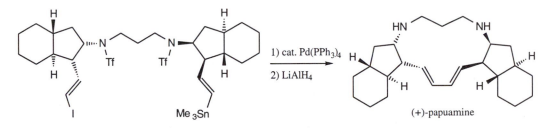

(+)-papuamine

Simple sulfonamide PGs are useful, though, in masking the nitrogen atom in aromatic heterocycles such as pyrrole, imidazole, and indole. Sulfonamide protection reduces the nucleophilicity of the heteroaromatic nitrogen, and the inductive electron-withdrawing effect of the sulfonyl group combines with the reduction in capacity for the nitrogen lone pair to contribute to the aromatic system to render the ring itself much less susceptible to electrophilic attack. In this respect N-protection also acts as a PG for the heterocyclic ring.

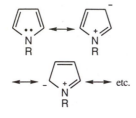

Because the nitrogen lone pair participates in the aromatic π-system, nucleophilic attack at sulfur is much more facile (and the amine is a more effective leaving group) than in *N*-sulfonyl *alkyl* amines. Thus, under mildly basic conditions, cleavage of both acetates, and the *p*-Ts- group from the indole nitrogen, proceeds without affecting the cycloheptyl sulfonamide (which is probably deprotonated under the reaction conditions).

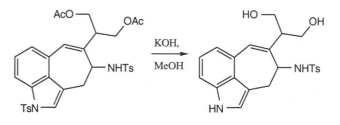

3.4 Deprotection by β-elimination

One of the mechanistic cornerstones of PG chemistry is the rapid ejection of an anionic leaving group from a position adjacent to a carbanion to form a carbon–carbon double bond. This *β-elimination* reaction is driven by the higher stability of a negative charge centred on a heteroatom compared with one centred on carbon, the enthalpic gain in generating an alkene in conjugation with a stabilising group (see below) and, in some cases, the enthalpic and entropic gain associated with the liberation of CO_2 or SO_2.

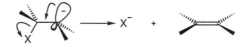

For this to be an efficient process the precursor carbanion has to be formed under conditions that are tolerated by other FGs in the molecule; the most obvious method—deprotonation—requires anion-stabilising functionality.

Anion stabilisation

Both inductive (σ-framework) and mesomeric (π-framework) electron withdrawal contributes to anion stability. Also of importance are: (1) the s-character of the orbital that formally holds the negative charge, (2) aromaticity (see below), and (3) anion overlap with aligned empty orbitals.

Carbanions are also stabilised by solvation, particularly by hydrogen-bond donors (protic solvents include H_2O, alcohols, AcOH, etc.).

Electrons in s-orbitals are more stabilised by the nucleus than those in p-orbitals; this is reflected in the pK_a values of the simplest hydrocarbons:

	pK_a		
HC≡CH	25	sp	(50% s)
$H_2C=CH_2$	36.5	sp^2	(33% s)
$CH_3–CH_3$	42	sp^3	(25% s)

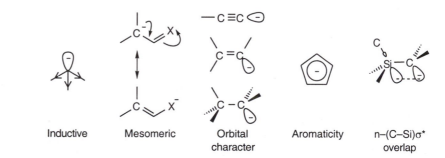

| Inductive | Mesomeric | Orbital character | Aromaticity | n–(C–Si)σ* overlap |

Deprotonation

Removal of a proton from the 9-position of fluorene generates a planar cyclic molecule containing 14 π-electrons, an aromatic system that satisfies Hückel's $4n + 2$ rule ($n = 3$), the central ring resembling the cyclopentadienyl anion which is isoelectronic with benzene (six π-electrons, $n = 1$).

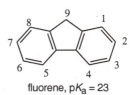

fluorene, $pK_a = 23$

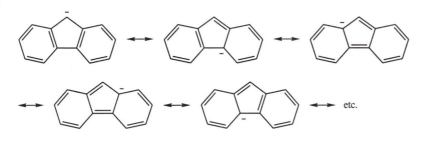

This stabilisation forms the basis of the 9-fluorenylmethyl (Fm) PG to protect thiols and carboxylic acids respectively as their Fm-thioethers and Fm-esters, RS$^-$ and RCO$_2^-$ being reasonable leaving groups (*cf.* Section 3.2).

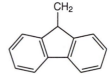

9-fluorenylmethyl (Fm)

$$R-SH \xrightarrow[i\text{-}Pr_2NEt]{Fm\text{-}Cl,} R-SFm$$

$$R\overset{O}{\underset{}{\|}}OH \xrightarrow[DCC, DMAP]{Fm\text{-}OH,} R\overset{O}{\underset{}{\|}}OFm$$

A 2°-amine (e.g. Et$_2$NH, piperidine, or morpholine) is used both to effect deprotection and to trap the resulting 9-methylene fluorene in an essentially irreversible addition of the amine that drives the deprotection to completion.

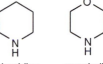

piperidine   morpholine

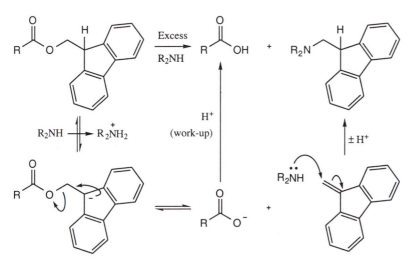

The leaving-group ability of RO$^-$ and R$_2$N$^-$ is insufficient to allow easy liberation of alcohols and amines by direct β-elimination from Fm-ethers and Fm-amines; therefore, these FGs require an interceding –CO$_2$– group [point (3), page 13] to confer base-lability. Fluorenylmethyloxycarbonyl (Fmoc) protection is carried out by treatment of an alcohol or amine with an Fmoc-X

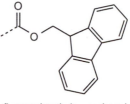

fluorenylmethyloxycarbonyl (Fmoc)

hydroxybenzotriazole (HOBt)

reagent (X = Cl, N$_3$, or OBt) in the presence of pyridine or NaHCO$_3$. When deprotection is required, treatment of the derived Fm-carbonates (RO-Fmoc) and Fm-carbamates (R$_2$N-Fmoc) with base ejects a carboxylate leaving group (*cf.* Fm-esters) but this time, either directly or when the carboxylate ion is protonated (in the reaction or during acidic work-up), loss of CO$_2$ yields the free alcohol or amine.

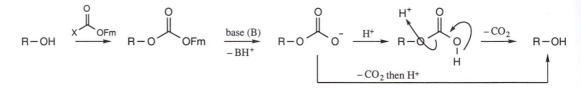

The Fmoc group is used widely in the N-protection of amino acids during solid-phase peptide synthesis since it is stable to the acidic conditions that cleave Boc groups from amines, is quite resistant to catalytic hydrogenolysis used in the deprotection of *N*-benzyl and *N*-Cbz groups, but can be released without affecting these other amine PGs with base as discussed above, e.g.

Cleavage of *N*-Fmoc in the presence of *N*-Cbz, and *N*-Boc PGs.

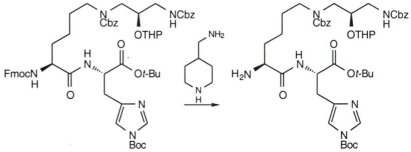

Although the use of a base-labile PG may be desirable (or unavoidable), carrying such a PG through a synthetic sequence imposes restrictions on the reagents that can be used for each of the steps. To cope with this situation base-labile PGs have been developed that can be introduced in a latent *unreactive* form, the reactivity to base only being unmasked when needed.

For example, 2-(methylthio)ethyl (Mte) esters are readily formed from carboxylic acids and 2-(methylthio)ethyl alcohol under acid catalysis. The sulfur atom in a thioether (RCH$_2$SMe) confers negligible acidity to the adjacent protons and the PG remains intact in the presence of most bases. When necessary, the acidity of these protons may be significantly increased by oxidation to the sulfone (RCH$_2$SO$_2$Me) or by S-alkylation (e.g. with MeI → RCH$_2$S$^+$Me$_2$) whereupon β-elimination is straightforward, e.g.

CH$_3$SCH$_3$	≈ 45
CH$_3$SOCH$_3$	35
CH$_3$SO$_2$CH$_3$	31.1
(CH$_3$)$_2$S$^+$CH$_3$	18.2

pK_a values (in DMSO) of sulfides, sulfoxides, sulfones, and sulfonium salts.

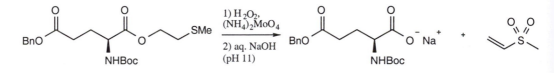

This concept is taken a step further in 1,3-dithianyl-2-methyl (Dim) esters which, after activation by oxidation to the *bis*-sulfone, are labile to deprotection under the mildest alkaline conditions.

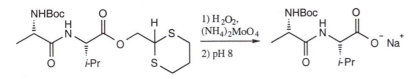

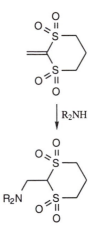

This idea has also been applied to protection of the *amino* terminus in peptide sequences but, as with the Fm- group, an interceding $-CO_2-$ group is required for efficient elimination. Furthermore, because the released free-base amine is significantly nucleophilic, the very electrophilic methylene *bis*-sulfone (formed during the elimination, see margin) has to be trapped by a sacrificial nucleophile and an excess of a 2°-amine (e.g. Me_2NH in MeOH) is used for Dim-carbamate (R_2N-Dmoc) deprotection.

Reductive methods

Metal–halogen exchange

A carbanion sited β to protected FGs may be generated under non-basic conditions by selective insertion of a metal into a carbon–halogen bond. This pathway operates in the selective deprotection of PGs built around the 2,2,2-trichloroethyl device (Section 1.4) for the protection of alcohols (trichloro-ethoxymethyl ethers), carboxylic acids (trichloroethyl esters), and carbonyl compounds [bis(trichloroethyl)acetals]. Following the same principles that interrelate the Fm and Fmoc PGs, the trichloroethyl group also occurs in trichloroethoxycarbonyl (Troc) PGs for alcohols, thiols, and amines. These PGs are usually removed by Zn in the presence of a proton source (H_2O, MeOH, or AcOH).

The trichloroethyl (Tce) device imparts lability to reduction.

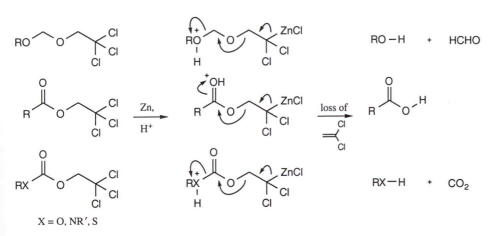

X = O, NR′, S

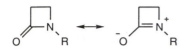

In β-lactams amide-type resonance does not operate to the extent that it does in normal amides because of the strain engendered in placing a double bond in a four-membered ring. Thus the carbonyl group is reactive towards addition of nucleophiles; this reactivity underlies the mode of action of penicillins and other β-lactam antibiotics.

Trichloroethyl PGs are introduced following standard etherification, esterification, and acylation protocols. They offer no special advantages over other ether, ester, and carbamate PGs (refer to Section 1.3) but, because Zn reacts only slowly with most FGs other than halides, deprotection is highly selective and sufficiently mild to leave other PGs intact. Simultaneous deprotection of two Tce-based PGs is achieved in the example below in which a hydrolytically sensitive β-lactam remains unscathed.

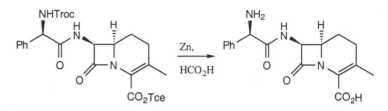

Acetals that incorporate special functionality to allow deprotection by, for example, reductive or basic pathways, are, nonetheless, still capable of cleavage by acidic hydrolysis.

The virtues of acetals as acid-labile PGs for carbonyl compounds have been discussed (page 31) but aldehydes and ketones protected as their bis(trichloroethyl)acetals can be deprotected in the presence of simple dialkyl acetals by exposure to Zn. Similarly, deprotection by Zn–halogen exchange and β-elimination combines with the advantages (ease of formation, stability) of cyclic acetals in 5,5-dibromo-1,3-dioxane and 4-(bromomethyl)-1,3-dioxolane acetals.

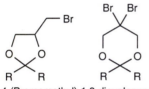

4-(Bromomethyl)-1,3-dioxolanes and 5,5-dibromo-1,3-dioxanes behave similarly.

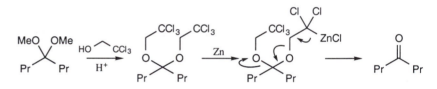

Reduction of π-systems

More prevalent than carbanion formation by metal–halogen exchange are β-elimination processes initiated by reduction of a C=O or C=C double bond. For example, phenacyl esters and phenacyl amines are moderately stable to acidic conditions yet are easily cleaved by reduction (Zn, AcOH; catalytic hydrogenolysis) as well as by exposure to soft nucleophiles (page 47).

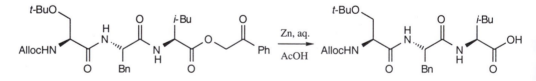

Formation of a ketyl radical-anion by electron transfer to a ketone.

Mechanistically, the metal donates two electrons overall to the phenyl ketone to produce a benzylic carbanion that initiates ejection of the protected FG as a leaving group (RCO_2^- or R_2N^-). The half-arrow notation in the first two steps of the scheme below (to produce a ketyl radical-anion then a benzylic carbanion) is used merely to keep track of the electron count. Overall, Zn(0) becomes oxidised to Zn(II).

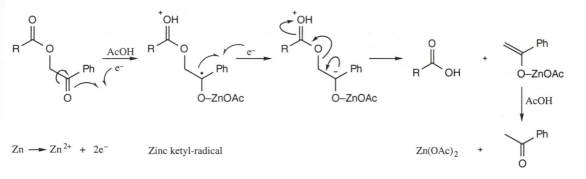

Zn ⟶ Zn²⁺ + 2e⁻   Zinc ketyl-radical

β-Elimination, following electron transfer to an aromatic system, forms a basis for deprotecting PGs containing benzyl functionality (PhCH$_2$–X). Whereas benzylic PGs are most often chosen because they have predictable lability towards catalytic hydrogenolysis (Chapter 5), they can also be cleaved with Na (or Li) in liquid NH$_3$ (often in the presence of a proton source such as *t*-BuOH). The benzyl group forms a key component of many PGs (page 89) but, in each case, the mechanism for dissolving-metal mediated deprotection follows the same general course:

(1) Addition of one electron   (→ radical anion, **A**)
(2) Ejection of the FG by β-elimination   (→ radical, **B**)
(3) Addition of a second electron   (→ anion, **C**)
(4) Protonation by NH$_3$ or ROH   (→ toluene)
(5) Further reduction of toluene may follow depending on the conditions.

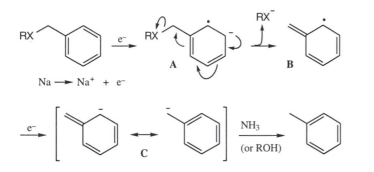

Na ⟶ Na⁺ + e⁻

The single electron in radical **B** and the electron pair in anion **C** are, of course, delocalised around the ring.

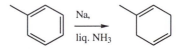

Further reduction of toluene (by two alternating electron transfers and protonations) results in 1-methyl-1,4-cyclohexadiene.

The radical-anion represented by structure **A** is just one of 18 interconvertible resonance isomers that imply an equal probability of the extra electron at any one carbon in the benzene ring:

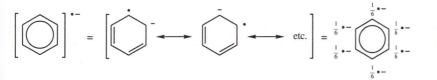

Various representations of benzene radical anion (*cf.* **A** above).

Fragmentation of siliconates

PGs based on the 2-(trimethylsilyl)ethyl unit undergo deprotection in a process analogous to the E1cB direct deprotonation pathway, a trimethylsilyl group taking the place of the proton. F⁻ ion, usually in the form of TBAF or CsF in a polar aprotic solvent (e.g. DMF), takes the place of the base. Thus, addition of F⁻ to the silicon atom produces a pentavalent siliconate intermediate which then fragments with ejection of the leaving group, the process being driven *inter alia* by the strength of the Si–F bond (page 74).

Bu₄N⁺F⁻   tetrabutylammonium
fluoride (TBAF)

Me₂NCHO   dimethylformamide
(DMF)

Elimination following siliconate fragmentation.

E1cB pathway; Y is an acidifying group.

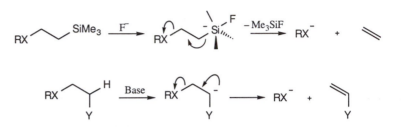

Protection of alcohols as 2-(trimethylsilyl)ethoxymethyl (SEM) ethers by treatment with SEM-chloride and Hünig's base (*i*-Pr₂NEt) provides levels of protection comparable to acetal PGs (page 25) but the deprotection conditions are similiar to those used for the removal of silyl ether PGs (Section 4.2). 2-(Trimethylsilyl)ethyl carbonates (Tmsec derivatives) are cleaved similarly but impart deactivation of the nucleophilicity of the protected oxygen. 2-(Trimethylsilyl)ethyl esters, derived from carboxylic acids under standard DCC/DMAP conditions, are much more stable than silyl esters but may still be liberated selectively with F⁻ ion.

F⁻ cleavage of SEM ethers is usually more difficult than that of silyl ethers because SEM ethers do not benefit from the electron withdrawing inductive effect of a directly attached oxygen atom (see mechanistic discussion in Chapter 4).

Trimethylsilylethyl PGs for OH protection.

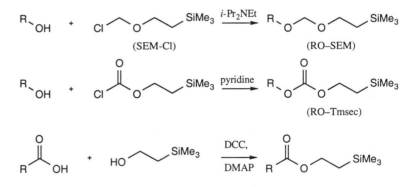

Amines protected either as 2-(trimethylsilyl)ethyl carbamates (Teoc derivatives, = Tmsec) or as 2-(trimethylsilyl)ethylsulfonamides (SES derivatives) benefit from the usual effects of carbamate or sulfonamide protection but may be deprotected under mild, non-reductive conditions with TBAF or CsF in an appropriate solvent. The selectivity in deprotecting SES-amines is shown in example (1) where four other amine PGs remain unaffected and example (2) provides an unusual application of the Teoc group in raising the efficiency of the Curtius rearrangement (R depicted in margin opposite).

Although Teoc-amines are usually produced from the amine with trimethylsilylethyl chloroformate (as for Tmsec-alcohols) they can also be prepared directly from benzyl amines (page 38).

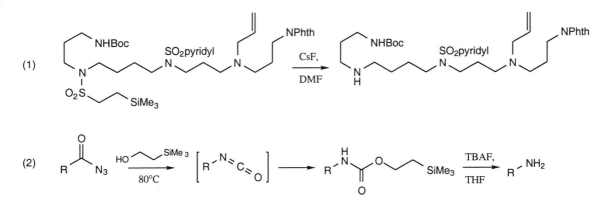

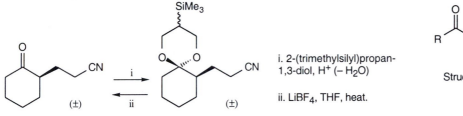

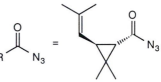

The trimethylsilylethyl device also occurs in *cyclo*-SEM acetal protection of ketones, e.g.

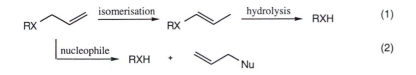

i. 2-(trimethylsilyl)propan-1,3-diol, H$^+$ (– H$_2$O)

ii. LiBF$_4$, THF, heat.

Structure for example (2) above.

3.5 Allylic protecting groups

Introduction

Selective and reversible coordination of a transition metal catalyst to an alkene results in a π-complex which, depending on the nature of any allylic substituents (RX), can lead either to (1) isomerisation of the alkene or (2) ejection of RX$^-$. Both of these outcomes form the basis of deprotection of PGs containing the allyl group. Allylic PGs are included in our discussion of base and nucleophile mediated deprotection because the isomerisation pathway (1) can also be effected by base and the displacement pathway (2) can be effected with a range of nucleophiles in an overall S$_N$2′ process.

Fe	Co	Ni	Cu
Ru	**Rh**	**Pd**	Ag
Os	**Ir**	Pt	Au

Important transition metals (bold) in allylic deprotection.

RX ⟶ (isomerisation) RX ⟶ (hydrolysis) RXH (1)

⟶ (nucleophile) RXH + Nu (2)

Allyl ethers, allyl amines, and allylic acetals

Where RX$^-$ is a poor leaving group, coordination of a transition metal centre to the alkene can be followed by reversible oxidative addition into the allylic C–H bond giving a metal hydrido π-allyl complex. Reductive elimination in the other direction leads to a new π-complex in which the alkene has migrated. Because these processes are freely reversible, the more stable alkene (under the reaction conditions) will predominate once equilibrium has been reached.

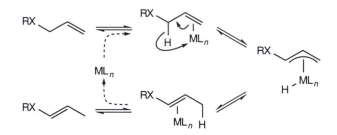

If, however, the transition metal bears a hydrogen atom an alternative isomerisation pathway can be followed that proceeds via a reversible hydride addition–hydride elimination pathway.

Complexes of Rh(I) (e.g. Wilkinson's catalyst, $(Ph_3P)_3RhCl$) and Ir(I) (e.g. $Ir(COD)(Ph_2MeP)_2PF_6$) are the usual choices.

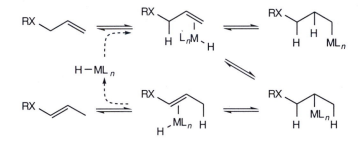

Whichever mechanism applies, if X = O or N, conjugation of the heteroatom lone electron pair(s) with the alkene provides stabilisation and a driving force to shift the equilibrium in the direction of the enol ether or enamine product.

Conjugation in (1) enol ethers and (2) enamines. A chemical consequence of this is raised electron density on the β-carbon of the alkene so that this site is readily attacked by electrophiles (e.g. H^+).

Subsequent hydrolysis of the resulting enol ether or enamine follows to complete the deprotection:

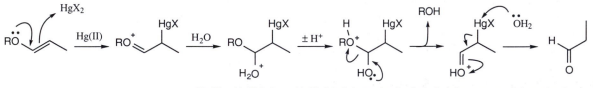

Hg(II) = $Hg(OAc)_2$ or $HgCl_2/HgO$ (enamine hydrolysis follows a parallel mechanism)

In cases where these conditions are not tolerated by the rest of the molecule, oxidative cleavage can be used. Thus, OsO_4 dihydroxylation of the enol ether, catalysed by NMO (dotted arrows), generates an unstable hemiacetal intermediate that breaks down to liberate the alcohol as in the penultimate step of the aq. Hg(II) method.

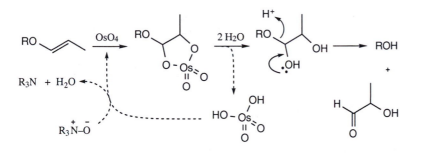

This is not applicable to enamines as the product amine may be oxidised further.

N-methylmorpholine-N-oxide (NMO)

Isomerisation of the alkene can also be achieved by strong base, an older method which is of most use in carbohydrate chemistry where the sugar nucleus is relatively insensitive to most bases. The two methods are complementary in their rates of isomerisation of differentially substituted allyl groups:

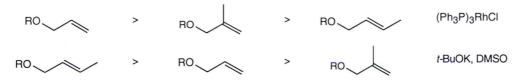

(Ph₃P)₃RhCl

t-BuOK, DMSO

Wherever this allylic ether arrangement occurs, it may, in principle, be susceptible to the same deprotection chemistry and the allyl group can therefore be viewed as another PG device. For example, acetals generated from carbonyl compounds and 2-methylenepropan-1,3-diol can be isomerised with Wilkinson's catalyst, then hydrolysed under very mild conditions that leave normal acetals intact.

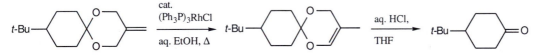

Allyl carbonates, esters, and carbamates

Where RX^- is an effective leaving group, oxidative addition occurs preferentially into the C–X bond forming a π-allyl complex, the metal being associated with or bonded to the leaving group. This process is reversible but the reverse process is inhibited, releasing the FG (as RXH), if an external nucleophile is present that is itself a poor leaving group. The external nucleophile can attack the π-allyl complex directly (overleaf, dotted arrow) or via addition to the metal with subsequent reductive elimination. Overall, RX^- should be a reasonable leaving group but poorly nucleophilic, and Nu^- should be a comparatively poor leaving group but a good nucleophile.

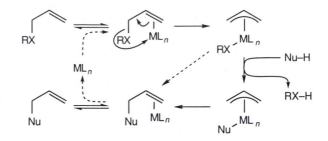

Soft nucleophiles tend to add directly to the allyl ligand (dotted arrow), hard nucleophiles add by prior reaction at the metal (solid arrow). (See page 36 for definitions of 'soft' and 'hard'.)

Nucleophiles for trapping the π-allyl complex include:

Bu₃SnH (transfers H⁻)

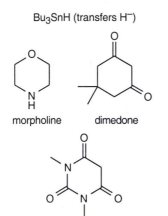

morpholine   dimedone

dimethyl barbituric acid

As carboxylates (RX⁻ = RCO₂⁻) are effective leaving groups in this reaction, allyl esters may be deprotected selectively in the presence of other alkyl esters; usually the catalyst is a Pd(0) complex [e.g. Pd(PPh₃)₄] and the added nucleophile a 2°-amine (morpholine or pyrrolidine) or the solvent (e.g. aq. EtOH), e.g.

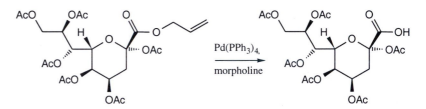

Representing the sequence of complexation and ejection of carboxylate by a truncated mechanism highlights the overall S$_N$2′ nature of the process and facilitates comparisons with other nucleophiles (notably organocuprates) that achieve the same result. The scheme below illustrates a plausible reaction pathway for the cleavage of an allyl ester with LiMe₂Cu, a soft source of Me⁻, and highlights the changes in oxidation state attending the metal centre in each case.

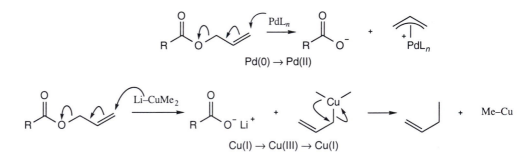

The fact that carboxylate is an effective leaving group in this chemistry has led to its combination with loss of CO₂ from monoalkyl carbonates and carbamates in the allyloxycarbonyl (Alloc) PG for alcohols and amines.

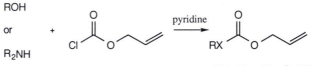

This PG has the usual advantages (compared with direct allylation) of ease of introduction (acylation with allyl chloroformate, rather than alkylation with allyl bromide), deactivation of the nucleophilic properties of the protected atom, and reliable cleavage in a single step with Pd(0) catalysis. As with allyl esters, an external nucleophile must be present during the deprotection step to prevent the liberated alcohol (or amine) from adding to the intermediate π-allyl complex. Deliberate omission of an external nucleophile allows a useful interchange of PGs (page 86) from RX-Alloc → RX-allyl as an alternative to direct allylation.

For example:

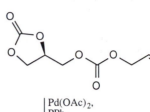

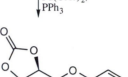

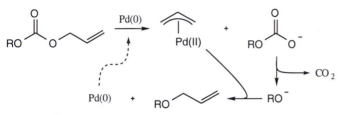

Further reading

Cited Oxford Chemistry Primer: #7: J. Jones, *Amino Acid and Peptide Synthesis.*
Ester deprotection: C. J. Salomon, E. G. Mata, and O. A. Mascaretti, *Tetrahedron,* **1993,** *49,* 3691.
Enzymatic deprotection: C.-H. Wong and G. M. Whitesides, *Enzymes in Synthetic Organic Chemistry,* Pergamon, Oxford, 1994; H. Waldmann and D. Sebastian, *Chem. Rev.,* **1994,** *94,* 911; M. Ohno and M. Otsuka, *Org. React.,* **1989,** *37,* 1 (use of pig liver esterase).
Allylic PGs: F. Guibe, *Tetrahedron,* **1997,** *53,* 13509 & **1998,** *54,* 2967.

4 Silyl protecting groups

4.1 Introduction

PGs based on the reactivity of silicon are also discussed in Chapter 1 and Section 3.4.

To categorise silicon PGs as either acid or base labile is impractical because both types of cleavage are possible and find routine use. Organosilicon chemistry is characterised by a high affinity for oxygen, which has led to the widespread use of silyl ethers for the protection of alcohols, and a (higher) affinity for fluorine, which provides a very selective deprotection pathway (F⁻ ion, see Section 4.6).

Bond strengths (kJ mol⁻¹)	Si–H	340	Si–C	320	Si–N	320	Si–O	530	Si–F	810
	C–H	420	C–C	335	C–N	335	C–O	340	C–F	450
Bond lengths (Å)	Si–H	1.48	Si–C	1.89	Si–N	1.74	Si–O	1.63	Si–F	1.57
	C–H	1.09	C–C	1.54	C–N	1.47	C–O	1.43	C–F	1.38

4.2 Alcohols

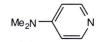

4-(dimethylamino)pyridine
(DMAP)

imidazole

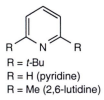

R = t-Bu
R = H (pyridine)
R = Me (2,6-lutidine)

Treatment of an alcohol with a trialkylsilyl chloride and an amine base (to scavenge liberated HCl) results in the formation of its silyl ether derivative. When the OH group is sterically hindered or the silane contains bulky alkyl groups a nucleophilic catalyst (imidazole or DMAP) is usually required for silylation to proceed at a convenient rate. In difficult cases, e.g. 3°-alcohols, trialkylsilyl trifluoromethanesulfonates can be used, although these very activated electrophiles can induce unexpected side-reactions.

$$ R{-}O{-}H \ + \ R'_3Si{-}X \ \xrightarrow{\text{base}} \ R{-}O{-}SiR'_3 $$

X = Cl   base = pyridine or i-Pr$_2$NEt or Et$_3$N, DMAP
X = SO$_2$CF$_3$ base = 2,6-dialkylpyridine (see margin)

Selectivity in alcohol silylation parallels that for the introduction of other PGs, i.e. 1° > 2° > 3°, although high levels of selectivity can only be relied upon when the bulkier silylating agents are used. Although the need for selectivity between different OH groups can be significant in guiding the choice from among the many silyl PGs available, more often the degree of protection is decisive.

Silylation replaces the weakly acidic OH proton that can destroy organometallic reagents but, for a given alcohol precursor, silyl ether cleavage becomes more difficult as the silyl alkyl substituents become bulkier.* Furthermore, for a given silyl PG, the more sterically encumbered

*A consequence of the mechanism of substitution at silicon (see below).

the alcohol, the more stable the derived silyl ether. This general view needs some qualification because the steric and electronic character of the alkyl substituents act differently depending on whether acidic or basic conditions are used for the deprotection. The key features are:

(i)   Increasing steric bulk raises stability.
(ii) Electron-withdrawing groups on silicon increase *acid* stability and decrease *base* stability, and vice versa.
(iii) Ease of cleavage with F^- parallels the ease of basic hydrolysis.

These features result in an order of stability for a given alcohol under either acid- or base-catalysed solvolysis conditions (a numerical stability rating is given for the silyl ethers that will be discussed individually below).

Me_3Si (1) ≈ $PhMe_2Si$ ≈ Ph_2MeSi < Et_3Si (64) ≈ $i\text{-}PrMe_2Si$ ≈ Pr_3Si ≈ Bu_3Si < Ph_3Si (400) < $i\text{-}Pr_2MeSi$ < $t\text{-}BuMe_2Si$ (2×10^4) < $Me_2(t\text{-hexyl})Si$ < $i\text{-}Pr_3Si$ (7×10^5) < $t\text{-}BuPh_2Si$ (5×10^6) < $TrMe_2Si$ < $t\text{-}Bu_2MeSi$

Acid-catalysed solvolysis (e.g. H^+, MeOH)

Me_3Si (1) ≈ $PhMe_2Si$ ≈ Ph_2MeSi ≈ Ph_3Si (1) < $i\text{-}PrMe_2Si$ ≈ Et_3Si (1.3×10^3) < Pr_3Si < Bu_3Si < $i\text{-}Pr_2MeSi$ < $t\text{-}BuMe_2Si$ (2×10^4) ≈ $t\text{-}BuPh_2Si$ (2×10^4) < $Me_2(t\text{-hexyl})Si$ < $i\text{-}Pr_3Si$ (10^5) < $t\text{-}Bu_2MeSi$

Base-catalysed solvolysis (e.g. HO^-, MeOH)

Phenyl (as opposed to methyl) substituents on silicon markedly decrease the stability towards base even though phenyl is effectively larger than methyl.* This supports a two-step mechanism of substitution in which (1) the attacking reagent adds to the silicon atom forming a pentavalent siliconate intermediate and then (2) the leaving group is ejected (the mechanistic opposite of S_N1 where the nucleophile adds *after* the leaving group has left).

*A-values (ΔG between axial and equatorial conformers of mono-substituted cyclohexanes) are used as a measure of the size of a group (smaller A-vaue ⇒ smaller group).

For Me: $A = 7.3$ kJ mol^{-1}
For Ph: $A = 11.7$ kJ mol^{-1}

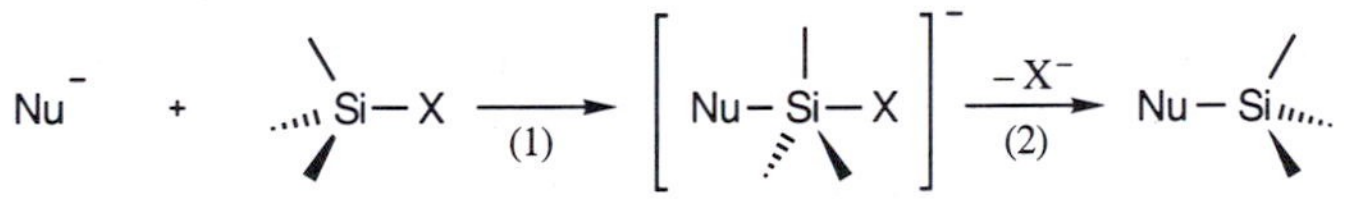

Simplified model for nucleophilic substitution at silicon via a siliconate intermediate.

The rate-determining first step is expected to be favoured by electron-withdrawing groups since these would disperse the negative charge nominally centred on silicon thus stabilising the siliconate intermediate and the TS that precedes it. Phenyl acts as an electron-withdrawing group in this regard and the lists above show that Me_3Si- (TMS-) and Ph_3Si- (TPS-) ethers have comparable stability towards base-catalysed solvolysis.

This mechanism is at best a simplification; although pentavalent siliconate *intermediates* are important during Si–C bond cleavage, a general base-catalysed solvolysis by a *concerted* S_N2–Si pathway is more likely in the base-mediated cleavage of Si–O bonds. However, the lifetime of the siliconate (intermediate or transition state) is of little practical importance as both pathways should be favoured by electron-withdrawing groups both on the silicon and in the proximity of the liberated OH group.

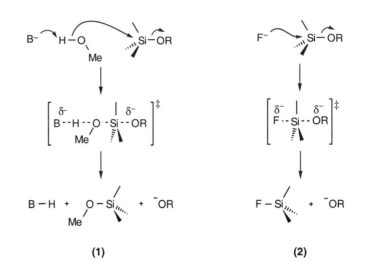

(1) General base-catalysed methanolysis of a silyl ether.

(2) Analogous process for F^- ion-induced desilylation in an aprotic environment.

(1)   **(2)**

Under *acidic* conditions, where the rate of reaction is influenced by the basicity of the alcohol oxygen, TPS is *ca.* 400× more stable than TMS. Cleavage is initiated by protonation and the electron-withdrawing phenyl groups disfavour the protonated form of the silyl ether.

RO–TMS

$Bu_4N^+F^-$ tetra-*n*-butylammonium fluoride (TBAF)

Alcohols derivatised as TMS-ethers are volatile and can be easily purified by distillation or analysed by gas chromatography (GC).

The trimethylsilyl (TMS) group is the simplest silyl PG for alcohols, being easy to introduce and rapidly removed with F^- ion (as TBAF/THF, KF/DMF, or aq. H_2SiF_6), or solvolysis (MeOH or EtOH) catalysed by either acid or base as discussed above. The TMS group provides only limited protection of alcohols against oxidation (see below) and for most synthetic purposes the TMS group is hydrolysed too easily to be practical, bulkier silyl ethers being preferred. An important application of the TMS group is in *temporary* protection (Section 1.5).

RO–TES

Compared with TMS the triethylsilyl (TES) group is hydrolysed around 50–1000× more slowly and is compatible with chromatography (SiO_2 is usual). Its main virtue is its ease of removal in the presence of bulkier silyl ethers; a 2% solution of HF in acetonitrile (CH_3CN) cleaves TES ethers in the presence of TBS ethers, e.g.

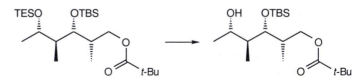

Often a free OH group is exposed as a prelude to oxidation. An advantage of the TES (and TMS) PG is that routine procedures for alcohol oxidation (e.g. Swern or CrO_3/pyridine) are effective on the silyl ether itself;

deprotection accompanies oxidation. This is a reliable procedure for protected 1°-alcohols in particular and, because these reactions do not work with TBS ethers, discrimination between differently silylated alcohols is straightforward:

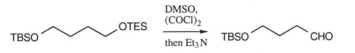

The TMS and TES groups are small enough that the ether oxygen atom is not hindered from attacking reactive electrophiles such as the chlorosulfonium salt formed in the Swern oxidation. Larger alkyl substituents provide a greater steric encumbrance and TBS (and higher) ethers are inert to these conditions.

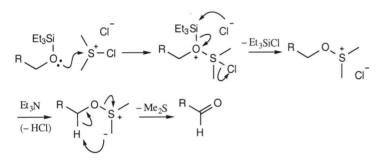

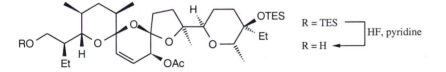

Selective cleavage of TES ethers in different steric and electronic environments is also possible:

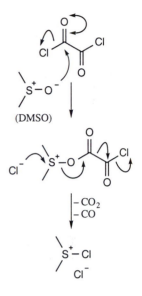

Formation of the oxidising agent (Me_2S^+–Cl) in the Swern oxidation.

RO–TPS

The triphenylsilyl (TPS) group has the advantages of dual reactivity—very easily removed under basic conditions ($\approx$ TMS) but lying between TES and TBS under acidic conditions—and steric bulk; the latter is useful in providing a basis for diasterocontrol (although it is less effective than TBS for this).

R = H		1.1:1
R = TPS		4:1
R = TBS		8.2:1

RO–TBS (a.k.a. RO–TBDMS)

The most generally useful silyl PG for alcohols is the *t*-butyldimethylsilyl (TBS) derivative, usually prepared using TBSCl in DMF with imidazole added as a nucleophilic catalyst (imidazole is more nucleophilic than most alcohols and, in the protonated form, is a superior leaving group to Cl⁻):

The abbreviations TBS and TBDMS can be used interchangeably.

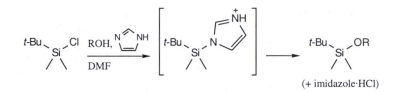

(+ imidazole·HCl)

Under these conditions (or Et$_3$N/DMAP in CH$_2$Cl$_2$) 1°-alcohols may be silylated in preference to 2°-alcohols, and 3°-alcohols remain unaffected. Introduction of a TBS group onto a 3°-alcohol usually requires TBSOTf in the presence of a pyridine base such as 2,6-lutidine (page 74).

TBS ethers are stable to a wide range of reaction conditions including organometallics, most metal hydrides, mild acid and base, chromatography, and most oxidising agents. They are, however, cleaved with aq. AcOH, ArSO$_3$H in MeOH, or the various sources of F$^-$ ion listed above. Fluorosilicic acid (H$_2$SiF$_6$) in aq. CH$_3$CN cleaves TBS (and other silyl) ethers more rapidly than HF in CH$_3$CN but its reactivity can be moderated by adding *t*-BuOH to enable selective cleavage. Mechanistically, H$_2$SiF$_6$ acts as a Lewis acid for the silyl ether which is then cleaved by water present in the solution; only catalytic quantities of H$_2$SiF$_6$ are required.

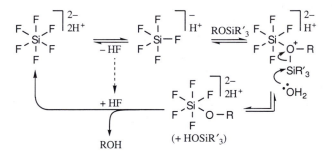

These catalytic cleavage conditions can lead to good selectivity for the deprotection of TBS ethers in the presence of bulkier silyl ethers (example 1) and between TBS derivatives of 2°- and 3°-alcohols (example 2).

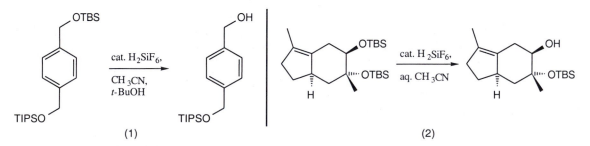

(1)   (2)

RO–TBDPS and RO–TIPS

<table>
<tr><td>

Under acidic solvolysis conditions TBDPS ethers are *ca.* 250× more stable than TBS ethers and *ca.* 10× more stable than TIPS ethers.

</td><td>

The *tert*-butyldiphenylsilyl (TBDPS) and tri-isopropylsilyl (TIPS) groups are more extreme versions of TBS. TBDPS ethers are the most tolerant of acidic conditions but the phenyl substituents lead to a rate of cleavage similar to

</td></tr>
</table>

that for TBS ethers under basic conditions. The bulkiness of this group facilitates selective protection of less-hindered OH groups, and its subsequent cleavage is straightforward following the general principles outlined above.

TIPS derivatisation of an alcohol confers the highest stability to base of the silyl groups in routine use and most nucleophilic and basic conditions are tolerated. The Further reading list should be consulted for a comprehensive overview of the TIPS group in synthesis.

4.3 Diols

Silicon protection of diols forms the third alternative after acetals and cyclic carbonates but the presence of the second oxygen atom bound to silicon makes these derivatives inconveniently labile to hydrolysis, especially under basic conditions. The only example of general use is the di-*t*-butylsilylene derivative as the bulk of the *tert*-butyl groups impedes nucleophilic attack to the extent that these derivatives usually survive chromatography.

Bonds to silicon are generally longer than those to carbon (by *ca.* 25–35%) (page 74) and accommodating these long bonds in rings (where the other atoms are C, N, or O) introduces angle strain. As a result five-membered silacycles are significantly more reactive than six-membered (and larger) analogues towards ring-cleavage by nucleophiles. Hence, silylene derivatives of 1,2-diols are more easily deprotected than those of 1,3- and 1,4-diols, the latter being stable over a wide pH range (4–10) at room temperature.

Whereas silylenes of 1,3-diols can be deprotected 'globally' alongside other silyl PGs (see cytovaricin, page 3), their lability allows selective deprotection in the presence of more resistant silicon-based PGs, as shown in the example where the nucleoside uridine is uniquely protected at the 2'-OH after silylene protection to block the 3',5'-diol.

di-*t*-butylsilylene

Calculated internal bond angles show the distortion from tetrahedral at the Si atom.

The example shows that silylenes are complementary to diol protection by acetonides or cyclic carbonates which preferentially protect *cis*-1,2-diol pairings.

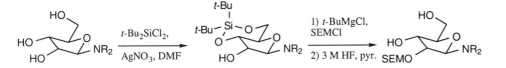

The double silylenes 1,1,3,3-tetra-*tert*-butyl- and 1,1,3,3-tetraisopropyl-disiloxanylidene (TIPDS) PGs can be used to protect the 3'- and 5'-hydroxyls of nucleosides and either the 3,4- or 4,6-diol pairings in carbohydrates as shown below. These derivatives form *medium* rings with 1,2- and 1,3-diols and therefore do not suffer from significant angle strain. The result of this is that smaller alkyl groups are acceptable, *i*-Pr being the usual choice.

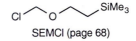

SEMCl (page 68)

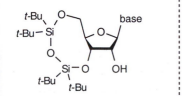

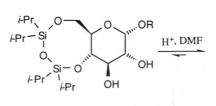

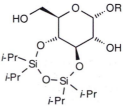

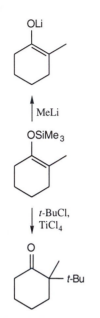

TMS enol ethers as enolate
equivalents.

4.4 Aldehydes and ketones

Silylation of carbonyl compounds (as their enolates) gives silyl enol ethers under conditions of kinetic (LDA, TMSCl, −78°C in THF) or thermodynamic control (Et₃N, TMSCl, heat in DMF). These derivatives, however, are little used as PGs *per se*, instead they are usually purified and then taken on in reactions where their utility as enolate equivalents is required (see OCP #1).

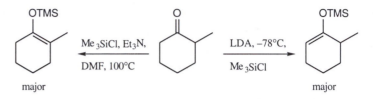

The most important silicon-based carbonyl PG—the *O*-trimethylsilyl cyanohydrin—prevents enolisation and nucleophilic attack at the carbonyl carbon. All the methods for forming these derivatives involve either TMSCl in combination with a source of CN⁻ (usually KCN), or TMSCN and a base (e.g. Et₃N). The reactivity of these adducts parallels that of TMS ethers, i.e. they are inherently unstable in the presence of nucleophiles and aqueous solutions but where there is more reactive functionality present they provide a useful alternative to, for example, acetal PGs.

Formation of the mono-silyl cyanohydrin derivative of quinones allows 1,2-addition of organolithium and Grignard reagents to only one of the two carbonyl groups, a process that proceeds very poorly without protection. Subsequent deprotection proceeds at room temperature in a few minutes to give the product of overall mono-addition.

TMS-cyanohydrins of aldehydes may be deprotonated (by LDA) and alkylated; deprotection affords a ketone, the cyanohydrin acting as an acyl anion equivalent.

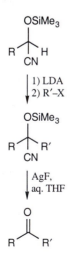

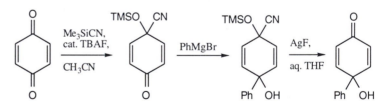

TMS-cyanohydrin formation with TMSCN only proceeds well if a small amount of an additive is present; PPh₃, F⁻, CN⁻, and amines are usually used. These additives probably form a pentavalent siliconate intermediate that acts as the true source of CN⁻, the silylating agent being generated *in situ*.

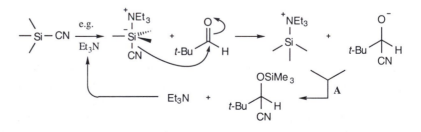

4.5 Amines

Refer to the table of bond strengths (page 74).

Step **A** in the previous scheme illustrates the ease of cleavage of Si–N bonds compared to Si–O bonds, this relative instability being reflected in the paucity of reliable silicon PGs for nitrogen compared with the huge range available for oxygen. Most of the amine PGs that *are* based on the reactivity of silicon keep the N and Si atoms separated (e.g. trimethylsilylethyl carbamate and trimethylsilylethyl sulfonamide, page 68).

The higher basicity of nitrogen compared to oxygen favours protonation and activation as a leaving group. Thus TBDPS derivatives of 1°-amines resist cleavage by bases, including hydroxide and alkoxide ions, but are decomposed within minutes under mildly acidic conditions. A corollary of this is that amides, lactams, and aromatic heterocycles can be usefully protected as their *N*-silyl derivatives since, being less basic, they are not as easily activated towards cleavage by protonation.

In the example below, TIPS-protection of pyrrole prevents competition by the nitrogen atom for electrophiles and (usefully) shifts electrophilic attack to the 3-position, the steric bulk of the isopropyl groups preventing efficient reaction at the 2-position (the usual site of attack, OCP #2).

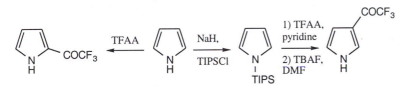

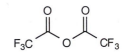

trifluoroacetic anhydride (TFAA)

For simple *alkyl* amines only one silyl PG is used with regularity: 'stabase' adducts are stable to organolithium reagents (below *ca.* −25°C), LDA, PDC, water, and aqueous solutions of NH_4Cl, $NaHCO_3$, and KF. They are cleaved with PCC and other acidic reagents (0.1 M HCl, 75% aq. AcOH) as well as moderately strong alkali (1.0 M KOH) and these latter conditions are often chosen for their deprotection. Stabase protection of glycine esters allows enolate alkylation to provide unnatural amino acids after deprotection.

tetramethylazadisilacyclopentane (H-stabase)

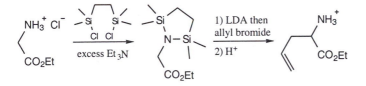

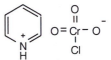

pyridinium chlorochromate (PCC)

4.6 Footnote

Deprotection of silyl PGs with F^- ion is normally highly selective but the basic properties of the hydrated ion in solution can interfere with functionality not containing silicon and can lead to unexpected reactions. Unfortunately, TBAF cannot be obtained in the anhydrous form as attempts to dry it eventually lead to decomposition to tributylamine; instead it is

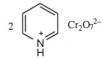

pyridinium dichromate (PDC)

obtained as the solid trihydrate or a 1.0 M solution in THF (with some water present). Various anhydrous sources of F⁻ have been suggested (e.g. SiF_4) but TBAF still remains the method of choice in most cases.

Further reading

Cited Oxford Chemistry Primers: #1: S. E. Thomas, *Organic Synthesis. The Roles of Boron and Silicon;* #2: D. T. Davies, *Aromatic Heterocyclic Chemistry.*

Reviews on silicon PGs and silylating agents: G. van Look, G. Simchen, and J. Heberle, *Silylating Agents,* Fluka Chemie AG, Buchs, 1995; M. Lalonde and T. H. Chan, *Synthesis,* **1985,** 817; J. Muzart, *Synthesis,* **1993,** 11 (oxidative deprotection); C. Rücker, *Chem. Rev.,* **1995,** *95,* 1009 (applications of the TIPS group); T. D. Nelson and R. D. Crouch, *Synthesis,* **1996,** 1031.

5 Redox deprotection

5.1 Oxidative methods

Oxidation-labile PGs offer a means of releasing FGs under essentially neutral conditions which is useful when hydrolytic deprotection is not tolerated and silyl PGs are unsuitable. In particular, a CH_2 group that links an FG to an electron-rich aromatic ring is prone to oxidation giving a hemiacetal (or equivalent), which is unstable with respect to aldehyde formation, and release of the FG (RXH in the diagram below; X = O, NH, etc.). Some oxidising agents (Ox) activate the CH_2 towards nucleophilic attack by water in a process equivalent overall to abstraction of hydride ion (H^-). This is favourable at benzylic sites because the resultant cation can be stabilised by delocalisation around the aromatic ring as well as by the adjacent heteroatom:

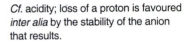

Cf. acidity; loss of a proton is favoured *inter alia* by the stability of the anion that results.

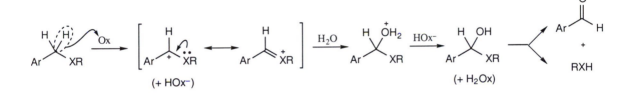

However, many reagents for methylene oxidation do not operate in this way and even those that *are* capable of direct hydride abstraction may, under some circumstances, achieve the same result by a more indirect route. Hydride removal requires the loss of the two electrons formally (and formerly) localised in a C–H bond; in many cases the electrons are lost not as a pair but singly and intermediate radical species become involved. Although this may seem complicated, the order of events usually follows a predictable pattern:

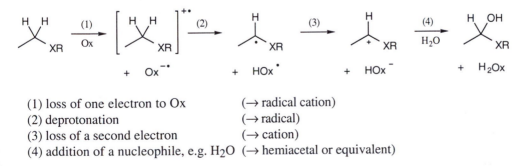

(1) loss of one electron to Ox   ($\rightarrow$ radical cation)
(2) deprotonation   ($\rightarrow$ radical)
(3) loss of a second electron   ($\rightarrow$ cation)
(4) addition of a nucleophile, e.g. H_2O ($\rightarrow$ hemiacetal or equivalent)

Note that Ox is reduced progressively to 'H_2Ox' as the oxidation of the substrate proceeds. When X is a heteroatom, fragmentation usually follows as

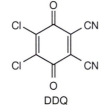

DDQ

CAN   (NH$_4$)$_2$Ce(NO$_3$)$_6$

Ph$_3$C$^+$ BF$_4^-$

depicted above but if X is a relatively poor leaving group, oxidation may proceed further to generate a carbonyl group at the methylene site.

Because the ease of hydride abstraction depends to a large extent on the stability of the resulting cation, and the ease of oxidation by single electron transfer is governed by the stability of the radical cation and radical intermediates, both mechanisms have a similar electronic requirement: electron-releasing functionality at the CH$_2$ position favours oxidation. Thus benzyl groups bearing electron-donating substituents are usually readily deprotected with oxidising agents (see margin for examples).

For example, *para*-methoxybenzyl (PMB) ether derivatives of alcohols combine effective mesomeric stabilisation with minimum inductive destabilisation. The table lists the times taken to release 2-phenylethanol protected with a range of methoxybenzyl substituents.

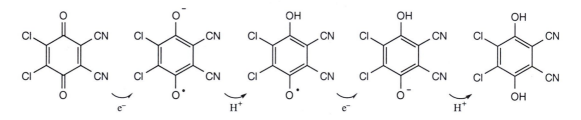

Ar–	Time (h)	Ar–	Time (h)
3,4-dimethoxyphenyl	<0.33	2-methoxyphenyl	3.5
4-methoxyphenyl	0.33	3,5-dimethoxyphenyl	8
2,3,4-trimethoxyphenyl	0.5	2,3-dimethoxyphenyl	12.5
3,4,5-trimethoxyphenyl	1	3-methoxyphenyl	24
2,5-dimethoxyphenyl	2.5	2,6-dimethoxyphenyl	27.5

The mechanism for this process follows the general stepwise electron transfer pathway outlined above; as the oxidation proceeds the DDQ is progressively reduced as shown:

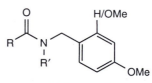

The electron-donating methoxy groups also enhance acid lability (page 22).

Methoxy-substituted benzyl groups are cleaved with sufficient reliability under oxidative conditions that they qualify as another PG device (Section 1.4). A few examples of the occurrence of this device in the protection of a range of FGs are given in the diagram below which also lists preferred deprotection reagents.

Alcohols   Amides

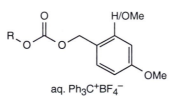

aq. Ph$_3$C$^+$BF$_4^-$ aq. CAN or aq. DDQ

Carboxylic acids

Diols

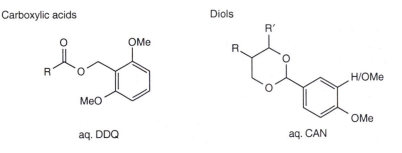

aq. DDQ

aq. CAN

Oxidising agents can activate PGs towards deprotection by analogy to addition of an electrophile; these cases are discussed separately (page 42).

5.2 Internal redox processes

Free-radical deprotection

The stabilising effect of aromatic rings on electron-deficient intermediates also underlies the free-radical deprotection of modified Bn- (and Tr-) ethers that incorporate a Br atom at the *ortho* position of (one of) the aromatic ring(s). This gives rise to a site for initiating a homolytic deprotection sequence, the product being the carbonyl compound derived from the protected alcohol precursor. Although *reductive* conditions are employed, the protected alcohol is *oxidised* during the deprotection (the bromobenzyl group is reduced).

Conventional deprotection of these modified benzylic PGs (Section 5.3) can be used if the alcohol itself is required.

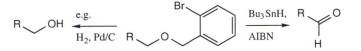

See also OCP #8.

After initiation, the reactive aryl radical abstracts a hydrogen atom to give an oxygen-stabilised radical (*cf.* page 87); this fragments to give the aldehyde and a relatively stable benzyl (or trityl) radical which continues the chain.

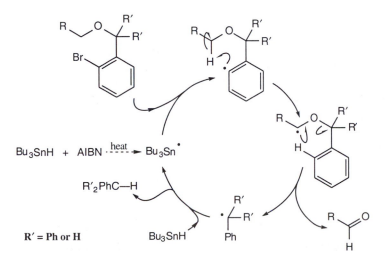

R′ = Ph or H

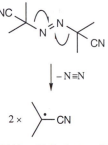

AIBN [azobis(isobutyronitrile)] decomposes with a half life of *ca.* 1.5 h at 80°C.

This method allows transformations to be achieved that would be more laborious using traditional methods. In the example shown, protection of the less-hindered 1°-OH group as a bromotrityl ether and then radical deprotection achieves overall oxidation of that OH group in the presence of the 2°-allylic OH (the latter would be oxidised more rapidly by most conventional oxidising agents).

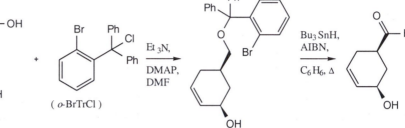

(*o*-BrTrCl)

Protecting group interchange (PGI)

Free-radical chemistry also plays a part in the *interchange* of PGs which is of importance in the selective release of just one FG in bifunctional molecules such as diols. Hydrolysis of cyclic acetal derivatives of diols (page 27) leads to release of both hydroxyls but if the steric or electronic environments of the hydroxyl groups are reasonably different then selective release of one of them may be possible. For example, treatment of benzylidene-protected carbohydrates with NBS (page 43) in CCl_4 promotes free-radical bromination of the acetal carbon leading to an unstable bromide which rearranges *in situ* to give a single bromoalkyl benzoate ester ('Hanessian procedure').

The selectivity in this conversion originates in the steps marked **A** and **B**. The mechanism of step **A** is analogous to that which operates in the allylic bromination of alkenes (Wohl–Ziegler bromination, OCP #35, page 49); thus initial H-atom abstraction by Br• is selective for the acetal position since this produces a carbon-centred radical stabilised by delocalisation into the

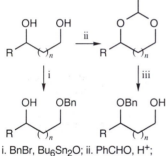

i. BnBr, Bu_6Sn_2O; ii. PhCHO, H^+; iii. *i*-Bu_2AlH (DIBAL). [$n = 0, 1$]

Direct mono-protection of a diol is usually selective for the less-hindered OH group. PGI via an acetal allows protection of the more hindered one.

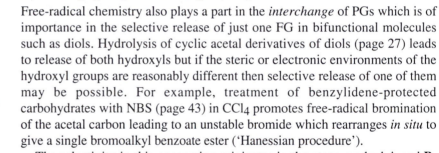

benzene ring and by the acetal O-atoms (see margin, opposite); bromination of this radical leads to the unstable intermediate. Secondly, in step **B**, attack by Br^- occurs at the sterically more available 1°-position to give the observed 1°-bromoalkyl benzoate. The mechanism for step **A** is shown in more detail as follows:

The Hanessian procedure can also be effected by slow addition of Br_2 in the presence of a radical initiator.

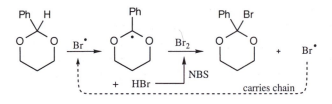

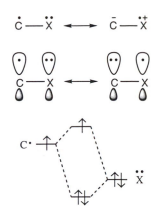

Representations of radical
stabilisation by an adjacent lone
pair of electrons.

Photochemical deprotection

Ultraviolet radiation possesses sufficient energy to promote an electron in a
bonding or a non-bonding molecular orbital to an unoccupied orbital resulting
in a molecular *excited state*. First formed as a singlet (electrons paired), the
molecule may undergo intersystem crossing (ISC, see OCP #39) to a triplet
state (the single electrons have the same spin); photochemistry can arise from
either the singlet or triplet state. This excitation is most efficient when the
energy of the incident radiation matches the energy gap between the affected
molecular orbitals (ΔE in the figure below). If functionality (a *chromophore*)
is present that absorbs in a specific, well-defined region, choice of radiation of
that wavelength will lead to selective excitation of that functionality and the
ensuing photochemistry should be controllable.

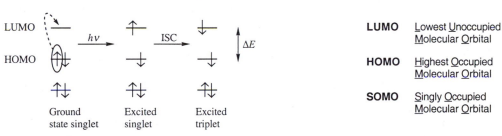

Ground state singlet	Excited singlet	Excited triplet

LUMO — Lowest Unoccupied Molecular Orbital

HOMO — Highest Occupied Molecular Orbital

SOMO — Singly Occupied Molecular Orbital

These principles combine with 1,5-hydrogen transfer and benzylic
activation in PGs built on the *o*-nitrobenzyl (ONB) group. These can be
introduced by methods analogous to those used for the unsubstituted benzyl
PGs, provide comparable levels of protection, but can be cleaved selectively
by photolysis.

The success of the deprotection step hinges on the rapidity with which an
oxygen radical abstracts a hydrogen atom from a position five-atoms distant.
This is a kinetically favourable process—the abstraction proceeding through a
six-membered cyclic transition state—and is energetically favourable since a
strong O–H bond is generated at the expense of a weaker C–H bond.
Ultimately, one of the oxygen atoms of the nitro group ends up attached to
the carbon that bears the protected FG (RX in the diagram overleaf) to
generate an unstable intermediate that breaks down rapidly to from *o*-nitroso-
benzaldehyde with liberation of the FG.

$\lambda_{\text{max.}}$ = 268.5 nm
$\varepsilon_{\text{max.}}$ = 7800

O–H   460–465 kJ mol⁻¹
C–H   400–420 kJ mol⁻¹
Bn–H   360 kJ mol⁻¹

Average bond enthalpies.

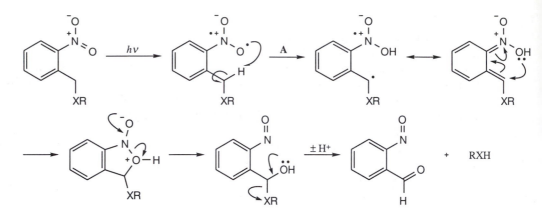

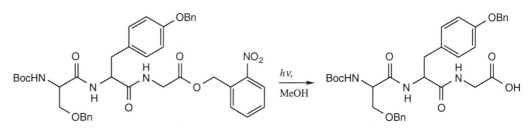

Step **A** could also be written as *proton* removal by -O⁻, the product would be the same.

An example of ONB deprotection from a carboxylic acid gives an idea of the range of FGs that remain unaffected by photolysis at 350 nm.

Hydroxyl groups may be protected as their ONB-ethers or ONB-carbonates, carboxylic acids as ONB-esters, and amines as ONB-amines or ONB-carbamates. In the case of ONB-carbonate and ONB-carbamate deprotection, the liberation of the alcohol or the amine includes a decarboxylation step as usual (see page 13).

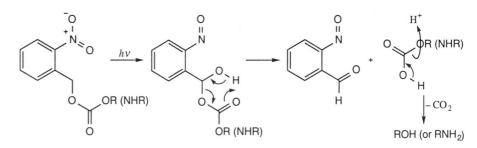

5.3 Reductive methods

Virtually all FGs show *some* reactivity towards reducing agents; therefore, for a PG to be deprotected selectively by reduction it must bear functionality that is particularly sensitive to exposure either to H_2, hydride transfer reagents, or electron sources.

Hydrogenolysis

PGs that incorporate a weak σ-bond or a benzyl group are prone to cleavage by exposure to H_2 in the presence of a metal catalyst, often Pd supported on carbon. Most FGs (alkenes and alkynes excepted) react with H_2 only slowly under these conditions so this method of deprotection can be very selective.

The sensitivity of benzyl PGs is associated with the relative weakness of benzylic (Bn–X) bonds compared with their alkyl–X counterparts (see margin) and the strong tendency of aromatic rings to form π-complexes with transition metals which brings the protected FG close to the catalyst surface. Although detailed mechanisms vary depending on the choice of catalyst, a working model for benzyl *ether* hydrogenolysis, catalysed by Pd/C, is outlined below (HA is usually a protic solvent, e.g. MeOH).

Hydrogenation is the act of adding H_2 to a molecule but, where addition results in cleavage of a σ-bond, the process is termed **hydrogenolysis**.

$$CH_3-H \quad \approx 410 \text{ kJ mol}^{-1}$$
$$PhCH_2-H \quad \approx 360 \text{ kJ mol}^{-1}$$

Comparison of alkyl–H and benzyl–H bond strengths.

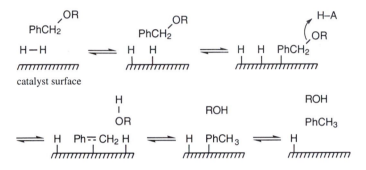

Key points: (1) reversible coordination of H_2 to the catalyst surface weakens the H–H bond leading to a coating of hydrogen in a reactive form; (2) reversible π-complexation of the benzyl group leads to bond weakening and cleavage to liberate the alcohol; (3) bond formation between hydrogen and the coordinated benzyl fragment generates toluene which diffuses away from the catalyst surface.

The benzyl group qualifies as a PG device because its incorporation into more complex PGs allows hydrogenolysis by the general mechanism outlined above, all benzylic–heteroatom bonds being replaced by bonds to hydrogen. Unsubstituted benzyl derivatives of alcohols, amines, and carboxylic acids provide effective protection against deprotonation and offer a degree of steric shielding. Electronic deactivation of nucleophilicity can be achieved by incorporating a $-CO_2-$ unit (page 13) as in the benzyloxycarbonyl (Cbz) group for amine protection in amino acid and peptide synthesis. The table below highlights the benzyl device as it occurs in a wide range of PGs cleaved by catalytic hydrogenolysis.

Thiol PGs incorporating Bn are not cleaved reliably by catalytic hydrogenolysis due to poisoning of the catalyst.

Cbz is also abbreviated as Z, after L. Zervas, its originator (1932).

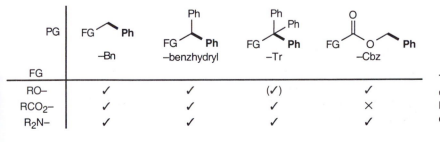

PG	FG⌒Ph −Bn	FG(Ph)(Ph) −benzhydryl	FG(Ph)(Ph)(Ph) −Tr	FG−CO−O⌒Ph −Cbz
FG				
RO–	✓	✓	(✓)	✓
RCO₂–	✓	✓	✓	✗
R₂N–	✓	✓	✓	✓

Trityl ethers are not reliably deprotected by catalytic hydrogenolysis; instead, mild acidic conditons are preferred (page 22).

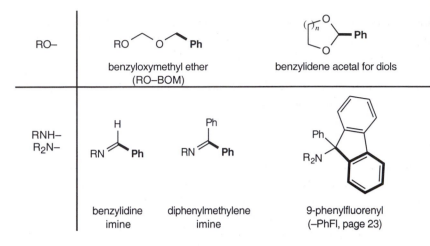

The occurrence of the benzyl device within a range of PGs for the same FG illustrates the general problem of choosing a PG from amongst the various forms that have similar deprotection requirements (H₂, cat. in this case). Working through an analysis of the choice of one of the four benzylic PGs for alcohols (Bn, benzhydryl, Cbz, BOM) will serve to outline the general treatment for other FG/PG combinations.

(a) How easily is each protecting group introduced? Each can be introduced readily but the ubiquity of Bn has led to a very wide range of conditions most of which involve treatment of the alcohol with a benzyl electrophile (BnCl or BnBr + cat. $Bu_4N^+I^-$) in the presence of a base [NaH; or Ag₂O in DMF (mild)]. If basic conditions cannot be tolerated then $BnOC(=NH)CCl_3$ + TfOH can be used, as can diazoalkanes provided a protic acid is present (particularly useful for benzhydryl protection).

Summary: PG selection requires a knowledge of:

(a) ease of protection
(b) level of protection
(c) ease of deprotection

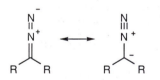

The right-hand canonical form is convenient for writing mechanisms.

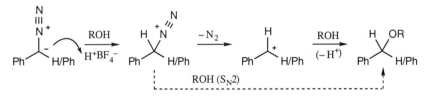

Whether S_N1 or S_N2 operates depends on the solvent, the alcohol, and the diazoalkane.

(b) What level of protection is offered? As mentioned above, benzyl and benzhydryl ethers provide no *electronic* deactivation of the oxygen atom's nucleophilicity but the extra phenyl group in the benzhydryl derivative provides a significant *steric* impediment to approaching electrophiles. Benzyl carbonate (Cbz) protection, on the other hand, brings the oxygen lone pairs into conjugation with the carbonyl group thereby protecting the oxygen from attack by electrophilic reagents.

(c) Are there exploitable differences in available cleavage methods? All four PGs are removable by catalytic hydrogenolysis but benzhydryl and BOM

ethers may also be removed with acidic treatment (*cf.* page 25); many deprotection options are available for Bn ethers, Lewis acid + nucleophile combinations being particularly useful (Section 2.3).

Careful consideration of each of these features in turn can aid the selection of the most suitable PG at the outset.

Benzyl amines

In preparing Bn derivatives of 1°-amines, over-alkylation to give dibenzylated products can be a problem and in these cases reductive amination with benzaldehyde can be used instead. First the 1°-amine forms an imine with the aldehyde and, in a second step, the imine is reduced with a hydride donor (e.g. $NaBH_4$). $NaBH_3CN$ may also be used to effect the reduction but, being a weaker reducing agent, protonation of the imine is required to activate it towards addition of H^- and AcOH is usually added. This reagent combination is especially useful since aldehydes and ketones are not reactive towards $NaBH_3CN$ under *mildly* acidic conditions, the imine being more readily protonated and thus activated selectively. The amine, aldehyde, $NaBH_3CN$, and AcOH can be mixed together to achieve both steps in a single reaction.

Two *N*-benzyl substituents usually provide sufficient steric impediment to prevent trialkylation ($\rightarrow Bn_3RN^+X^-$).

The cyano group in $NaBH_3CN$ removes electron density from the B atom, reducing the tendency of the reagent to transfer H^-.

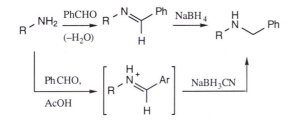

Named variants:

Wallach reaction
(HCO_2H used to reduce the iminium ion).

Leuckart reaction
(uses $HCO_2^- NH_4^+$).

Benzyl amines are less susceptible than benzyl ethers to catalytic hydrogenolysis but protonation speeds the process* and *transfer* hydrogenolysis also helps. Where neither of these modifications prove useful, stepwise deprotection via an intermediate carbamate can be used (page 38).

* The ease of hydrogenolysis is dependent on the leaving-group ability of the FG. Neutral R_2NH is a much better leaving group than R_2N^-.

Transfer hydrogenolysis:
H_2 is gained from another molecule, (e.g. cyclohexene, $HCO_2^- NH_4^+$, HCO_2H). Cbz carbonates are the most easily cleaved under these conditions.

Other reductive methods

Reductive deprotections mediated by dissolving-metals usually proceed by the formation of anionic intermediates that eject the FG by β-elimination, a general mechanistic pathway that is discussed in detail in Chapter 3. Alternatively, metals and other sources of electrons can deprotect PGs that incorporate an easily broken σ-bond, such as that between two heteroatoms.

The diagram overleaf presents a schematic molecular orbital picture for one-electron cleavage of a generalised σ-bond. Addition of an electron to a weak bond (O–O, N–O, S–S, etc.) gives a radical-anion (step **A**). The additional electron formally occupies the antibonding orbital (σ*) which leads to bond weakening and cleavage (step **B**) to produce a radical and an anion. The anion initially forms on the atom better able to support a negative charge but the radical component usually accepts a second electron; both fragments

HO–OH	215 kJ mol^{-1}
H_2N–NH_2	280 kJ mol^{-1}
HS–SH	275 kJ mol^{-1}

The bond energies in the alkylated counterparts of these structures are very variable but generally lower than the values listed above.

then react further or simply protonate.

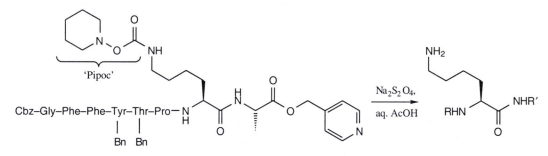

PGs that can be cleaved in this manner include disulfides (for thiol protection, page 6) and *N*-piperidinyl carbamates (for amine protection), the latter incorporating a $-CO_2-$ unit (page 13) for ease of synthesis. In most cases an acid is required to protonate the anionic intermediates and minimise side-reactions. Reducing agents other than metals can initiate single electron transfer processes (e.g. $Na_2S_2O_4$: $S_2O_4^{2-} \rightarrow 2SO_2 + 2e^-$), or achieve deprotection by donation of H^- (e.g. $NaBH_4$; $LiAlH_4$; RedAl). The example below illustrates discrimination between four potentially reducible PGs (Cbz, Bn, Pipoc, Pic) during the synthesis of a sequence of bovine insulin.

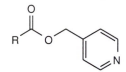

4-Picolyl esters (RCO_2Pic) can be cleaved by catalytic hydrogenolysis (*cf.* Bn esters) or hydrolysis; the basic N atom allows easy removal of the cleaved PG by acidification.

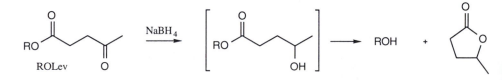

Gly glycine
Phe phenylalanine
Tyr tyrosine
Thr threonine
Pro proline

Amino acid abbreviations
(see OCP #7)

4-Azido- and 4-nitrobutyrate esters behave analogously (deprotection with aq. PPh_3 and Zn, HCl respectively).

PGs modified by incorporating readily reducible functionality are prone to deprotection under reductive conditions to which the unmodified analogues would normally be stable. For example, ester-protected alcohols and phenols are usually deprotected by either acid or base hydrolysis; achieving selectivity between them under these conditions is difficult. However, esters that contain a *latent* nucleophile that is only exposed on reduction are cleaved rapidly by cyclisation and ejection of the alcohol/phenol as a leaving group. For example, the ketone carbonyl group in 4-oxopentanoate (levulinate, Lev) esters may be reduced under mild conditions (e.g. $NaBH_4$), cyclisation giving a lactone and the deprotected alcohol.

These 'triggered' deprotections result in ester 'hydrolysis' under non-hydrolytic conditions that leave relatively reactive esters (e.g. methyl esters) intact.

3°-Amine oxides

3°-Amines may be protected *in situ* by protonation or complexation with BF$_3$ (page 14), the nitrogen lone electron pair becoming involved in bonding and therefore unable to compete for electrophiles. Oxidation (RCO$_3$H; H$_2$O$_2$) to the 3°-amine oxide achieves similar levels of protection and, when required, the 3°-amine can be regenerated reductively (H$_2$, cat; Zn/HCl; aq. SO$_2$).

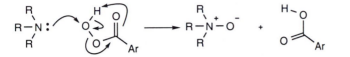

Peroxyacid oxidation of amines follows a course similar to alkene epoxidation.

Shown below is a synthesis of codeine from the MOM ether of morphine where protection of the 3°-amine prevents competing N-methylation.

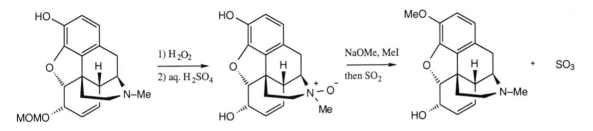

N-Sulfonamides

The synthesis of morphine alkaloids can be used to illustrate another set of reduction-labile nitrogen PGs, the *N*-sulfonamides. These derivatives are readily formed from 1°- and 2°-amines with the appropriate sulfonyl halide or anhydride and provide excellent protection against electrophilic attack by virtue of powerful electron withdrawal from the nitrogen atom.

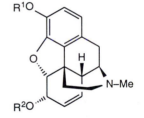

R^1 = R^2 = H, morphine
R^1 = Me, R^2 = H, codeine
R^1 = R^2 = Ac, heroin

The problem often lies in getting them off again when no longer required (but see page 62) as they are *so* stable (particularly towards acidic conditions) that vigorous cleavage conditions are usually required. However, sulfonamide cleavage by dissolving metal reduction—the usual method—can offer unexpected bonuses, as in the key step of another synthesis of a codeine analogue:

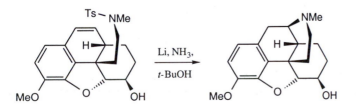

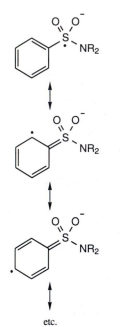

etc.

The extra electron in the first-formed radical anion is delocalised over the aromatic ring and the sulfonyl group.

The mechanism of reductive detosylation helps to explain this result. Electron transfer from lithium initiates cleavage of the N–S bond to generate *p*-toluenesulfinate anion (Ts$^-$) and a reactive nitrogen-centred radical that cyclises to form the piperidine ring.

cyclises in the above example

This example illustrates that a PG need not be merely a necessary evil but, if chosen correctly, with account taken of its reactivity and deprotection chemistry, may be actively involved in the key transformations that lead to molecular assembly.

Further reading

Cited Oxford Chemistry Primers: #7: J. Jones, *Amino Acid and Peptide Synthesis;* #8: C. J. Moody and G. H. Whitham, *Reactive Intermediates;* #35: G. D. Meakins, *Functional Groups: Characteristics and Interconversions;* #39: C. E. Wayne and R. P. Wayne, *Photochemistry.*

Free-radical and photochemical deprotection: D. P. Curran and H. Yu, *Synthesis,* **1992,** 123; V. N. R. Pillai, *Synthesis,* **1980,** 1.

Index

A-values 75
Acetals 7,12,15,24,25,27,31,43
  Acetonide 1,28,29
  Allylic 69,71
  Benzylidene 28,29,86
  Butane diacetals 30
  Chiral 33
  Cyclic 27,32,44,66
  cyclo-SEM 69
  Dimethyl 31
  Dithio- 40
  Ethylidene 29
  Methylene 29
Alcohols 3
  Silylation 74
Aldol reaction 6
Alkenes 9,10
  Isomerisation 70
Alkynes 10
Allylic groups 12,69
Ambident nucleophiles 53
Amides 5,44,49,57,58
  Acetamides 58
  Benzamides 58
  Formamides 58
  Phenyl acetamides 59
  Trifluoroacetamides 58
Amine oxides 14,93
Amines 5
  Allyl 69,70
  Benzyl 23,38,91
  Silylation 81
  Temporary protection 14
Ammonium salts 5
Anion stabilisation 46,62
$B_{Ac}2$ 49
$B_{Al}2$ 46
Benzylic groups 12,21,67,89
Benzylic oxidation 83,86
Boranes, oxidation 16
t-Butyl group 19
Carbamates 5,24,38,60
  Allyl 72
  Benzyl 19,24,89
  t-Butyl 20

1,3-Dithianyl-2-methyl 65
9-Fluorenylmethyl 63
p-Methoxybenzyl 24
Piperidinyl 92
2,2,2-Trichloroethyl 60,65
2-(Trimethylsilyl)ethyl 39,68
Carbonates
  Allyl 72,73
  Benzyl 89
  Cyclic 54
  9-Fluorenylmethyl 63
  2,2,2-Trichloroethyl 65
  2-(Trimethylsilyl)ethyl 68
Carbonyl compounds 6
Carboxylic acids 7
Cation scavengers 20
Cation stabilisation 18,23
Condensation 27
Curtius rearrangement 68
Decarboxylation 13
Di-*t*-butylsilylene 79
Diels–Alder reaction 10
Difunctional molecules 30
Dioxanes 28
Dioxolanes 28
Diphenylmethylene imines 24
Disiloxanylidenes 79
Dissolving metal reduction 67
E1cB 68
Electronegativity 46
β-Elimination 45,51,56,62
Enzymatic cleavage 51,56,58
Epoxide deoxygenation 9
Esters 46,50,55,56
  Acetate 2,50,51
  Allyl 21,71,72
  4-Azidobutyrate 92
  Benzoate 50,51
  Benzyl 8,21,47
  t-Butyl 8,19
  Chloroacetate 53
  Chloroalkyl 48
  Diisopropylmethyl 49
  1,3-Dithianyl-2-methyl 65
  9-Fluorenylmethyl 63

Formate 50
Levulinate 53,92
Mesitoyl 52
Methallyl 21
Methyl 46
Methylthioethyl 64
p-Nitrobenzoyl 52
4-Nitrobutyrate 92
Phenyl 8
Picolyl 92
Pivaloate 4,50,52
Silyl 16
2-(Trimethylsilyl)ethyl 68
Ethers 4
  Allyl 69,70,73
  Benzyl 2,44
  o-Bromobenzyl 85
  o-Bromotrityl 85
  t-Butyl 19
  t-Butyldimethylsilyl 3,22,77
  t-Butyldiphenylsilyl 14,55,78
  Diethylisopropylsilyl 3
  Dimethoxytrityl 22
  Ethoxyethyl 26
  Methoxybenzyl groups 84,85
  Methoxyethoxymethyl 37
  Methoxymethyl 25,37
  1-Methoxy-1-methylethyl 26
  Methyl 36
  Methylthiomethyl 39
  Monomethoxytrityl 22
  4-Pentenyloxymethyl 42
  Silyl 4
  Tetrahydropyranyl 26
  Triethylsilyl 3,76
  Triisopropylsilyl 53,78,81
  Trimethoxytrityl 22
  Trimethylsilyl 76
  2-(Trimethylsilyl)ethyl 43,68
  Triphenylmethyl 3,22,23
  Triphenylsilyl 77
  Trityl 3,22,23
Examples
  Artemisinin 16
  Bovine insulin 92

Calicheamicinone 15
Chalcogran 41
Codeine 37,93
Cytovaricin 3
Glucose 1,28
Meldrum's acid 31
MK-383 16
Morphine 37,93
Papuamine 61
Prostaglandin E$_2$ 52
Tirofiban 16
Fluorene 63
Fluoride ion 81
Free-radical deprotection 85
Functional groups 3
Gabriel synthesis 59
Hanessian procedure 86
Hard acids/bases 36
Heterocycle protection 61
Hydrogenolysis 89
Imides 58,59
Iodolactonisation 56
β-Lactams 66
Leuckart reaction 91
Metal–halogen exchange 65
Mitsunobu reaction 56,60
Mono-protection of diols 86
o-Nitrobenzyl group 13,87,88
Noyori acetalisation 15,32
Orthoesters 8,33

Trioxaadamantane 34
Trioxabicyclo[2.2.2]octane 8,34
Oxathiolanes 40
Oxidative deprotection 83
Peptides 8,57
Phenacyl group 47,66
9-Phenylfluorenyl group 23
Phenyl group, oxidation of 8
Photochemical deprotection 87
Phthalimides 49,58,59
Protecting groups
  Devices 12
  Graduated lability 3
  Interchange 86
  Nesting 49
  Orthogonal sets 2
  Requirements 2
  Selection 90
  Strategy 1
Quinone protection 80
Radical-anions 66,67,91,92
Radical-cations 83
Reductive amination 91
Reductive deprotection 65,88
Ring-size nomenclature 54
Schotten–Baumann conditions 5
Siliconate fragmentation 68
Silyl enol ethers 80
Silyl groups 12

Silylation mechanisms 75,76
Silylation rates 75
Silylation selectivity 74
Silylenes 79
Single electron transfer 42
Soft acids/bases 36
Spiroacetals 41
Sulfonamides 61,93,94
  Trifluoromethyl 61
  2-(Trimethylsilyl)ethyl 68
Sulfur
  Compounds 6,48
  Protecting groups 6,48,63,92
Sulfur-containing groups 39
Swern oxidation 77
Synchronicity 37,47
Temporary protection 13
  Carbonyl groups 14
  Hydroxyl groups 16
Transfer hydrogenolysis 91
Transition metal complexes 9,10,69,70
2,2,2-Trichloroethyl group 65
Triggered deprotection 92
Trimethylsilyl cyanohydrins 80
Urethanes, see *Carbamates*
von Braun reaction 38
Wallach reaction 91
Williamson ether synthesis 4,36

Abbreviations

Ac 50
ACE-Cl 38
AIBN 85
Alloc 72
BDA 30
Boc 20
Bn 2,12,85
Bz 50
CAN 42,84
Cbz 19,24,89
CSA 26
DBU 22,51
DCC 4,50
DCU 50
DDQ 42,84
DEAD 56

DEIPS 3
DIAD 60
DIBAL 23,53
Dim 65
DMAP 4,74
DMDO 14
DMF 8,46,68
Dmoc 65
DMSO 40,46
DMTr 22
EE 26
EEAC 52
FG 1
Fm 63
Fmoc 63
Fp 9

HOBt 64
KHMDS 23
LDA 5
Lev 53,92
MCPBA 14,55
MEM 37
Mes 52
MME 26
MMTr 22
MOM 25,37
Moz 24
Mte esters 64
MTM 39
NBS 42
NCS 42
NIS 42

NMO 71
OBO 8,34
ONB 13,87,88
PCC 81
PDC 1,55,81
PG 1
PGI 86
PhFl 23
Phth 58,59
Pic 92
PIFA 42
Pipoc 92
PLE 52
PMB 84
PNBz 52
POM 42

PPTS 26
Pv 4,50
Red-Al 61
SEM 42,68
SES 68
SET 42
Stabase 81
TBAF 39,68,76
TBDPS 14,55,78
TBS 3,22,77
Tce 65
Teoc 39,68
TES 3,76
Tf 15,32,48,61
TFA 8

TFAA 81
THP 26
TIPDS 79
TIPS 53,78,81
TMS 76
Tmsec 68
TMTr 22
TOA 34
TPS 77
Tr 3,22,23
Troc 65
Z 24,89